THE
REVERSE AGING
POWER OF EPITALON

Unlocking The Secrets
Of Longevity And
The Fountain
Of Youth

DR. JON HARMON, DC, BCN, ICP

DISCLAIMER

The information presented in this book has been compiled from my experience and research. It is offered as a view of the relationship between healthy living, exercise, balance and health. This book is not intended for self-diagnosis or treatment of disease, nor is it a substitute for the advice and care of a licensed healthcare provider. Sharing of the information in this book with the attending physician is highly desirable.

This book is intended solely to help you make better judgements concerning your long-term health goals. If you are experiencing health problems, you should consult a qualified physician immediately. Remember early examination and detection are important to successful treatment of all diseases.

TABLE OF CONTENTS

CHAPTER 1

AN INTRODUCTION TO EPITALON

The desire to stay "forever young" has been a universal yearning throughout human history. This quest for youthfulness, vitality, and longevity has shaped our cultural narratives, our scientific pursuits, and our individual behaviors in profound ways. Today, the field of longevity and anti-aging is witnessing a surge in novel and innovative therapies aimed at preserving our natural youth and life force for longer periods.

Over the past decade, the realm of anti-aging therapies has witnessed remarkable advancements. These developments, however, are not just focused on creating new solutions but also on understanding and harnessing the body's innate ability to heal, regenerate, and rejuvenate.

This shift in focus from external interventions to internal healing mechanisms marks a significant paradigm shift in the field—one that is both empowering and revolutionary. This paradigm shift is reflected in the growing body of research centered around substances like coenzyme Q10, curcumin, ginsenoside Rg1, bioidentical hormones, and geroprotectors, which are believed to slow down the aging process. Powerful herbs such as Ginkgo biloba, vinpocetine, and ginseng are also gaining traction, with their potential to boost brain power, increase mental energy, and enhance overall health.

Yet, the pursuit continues unabated. The future of anti-aging therapies may lie in reversing age-related changes once they have occurred or in decelerating or preventing these changes from the outset.

The industry is also witnessing a surge in longevity trends, with new therapies and interventions being explored and adopted. One such innovative therapy is the use of Epitalon.

Epitalon and Its Biochemical Composition

Epitalon, also known as Epithalone, is a synthetic tetrapeptide that has garnered significant scientific interest due to its potential anti-aging and longevity properties. The natural form produced by the pineal gland is called "Epithalamin." A tetrapeptide is a type of peptide composed of four amino acids linked together by peptide bonds. In the case of Epitalon, these specific amino acids are alanine, glutamic acid, aspartic acid, and glycine.

The sequence of these amino acids in Epitalon — alanine-glutamic acid-aspartic acid-glycine (Ala-Glu-Asp-Gly) — is not arbitrary. Instead, it has significant implications for the peptide's function and stability. This specific sequence allows Epitalon to interact with particular receptors and enzymes in the body, contributing to its biological effects. Among the most notable attributes of Epitalon is its potential to activate telomerase, an enzyme that repairs and extends the length of telomeres.

Telomeres are protective caps at the end of our chromosomes and the number of times a cell can divide is determined by the telomeres being able to hold the endings of the genes together. This limit is called the "Hayflick limit". Once the telomeres are too short to hold the double-helix gene structure together, the cell dies. The increase in telomerase production activated by Epitalon could effectively slow down this shortening process, thereby potentially slowing down the biological aging process.

In repairing the telomere endings, each cell is able to divide up to 30% more times before it reaches the Hayflick limit.

Interestingly, the sequence of these amino acids in Epitalon mirrors that found in Epithalamin, a peptide naturally produced in the pineal gland. The pineal gland is a small endocrine gland located in the brain that is responsible for the production of melatonin, a hormone that regulates sleep patterns.

The biochemical composition of Epitalon is represented as $C_{14}H_{22}N_4O_9$, indicating it contains:

· 14 carbon atoms,
· 22 hydrogen atoms,
· 4 nitrogen atoms,
· and 9 oxygen atoms

This molecular structure results from the unique arrangement and bonding of its constituent amino acids. Each of these amino acids plays a crucial role in the body.

Alanine

Alanine's role in protein synthesis could be significant in the context of Epitalon. The process of telomere extension involves the replication of DNA, a process requiring protein synthesis.

Alanine's involvement in glucose metabolism could also indirectly affect the energy required for the functioning of telomerase and the extension of telomeres.

Glutamic Acid

Glutamic acid, like alanine, plays a vital role in protein synthesis. This function could potentially contribute to the synthesis of telomerase, aiding in the extension of telomeres.

Additionally, glutamic acid's role as a neurotransmitter might have implications for brain health, a significant factor given that aging often impacts cognitive functions.

Aspartic Acid

Aspartic acid's contribution to protein construction and energy production may also intersect with the function of Epitalon. The citric acid cycle, in which aspartic acid is involved, is crucial for generating the energy needed for various cellular processes, possibly including the extension of telomeres.

Glycine

Finally, glycine, despite its simplicity, has a crucial role in the production of DNA, other essential compounds, and proteins. This function might be particularly relevant in the context of Epitalon, considering that the extension of telomeres involves DNA replication.

The Development of Epitalon

The history of Epitalon's discovery is intertwined with a larger narrative of scientific exploration and the quest to understand the mechanisms of aging. The pioneering work of Professor Vladimir Khavinson and his team was part of a broader effort to explore the role of the pineal gland and its secretions in aging processes. Starting in the 1970s, Khavinson, who was then serving as a military doctor in the Soviet Union, began exploring the potential of biologically active compounds isolated from animal pineal glands.

His interest in this area stemmed from earlier studies that had shown the pineal gland's role in regulating various physiological functions, including aging. The first major breakthrough came in the form of Epithalamin, an extract derived from the pineal glands of calves. Extensive experimentation on different animal models demonstrated that Epithalamin could extend lifespan and improve physical endurance, among other health benefits.

However, the use of Epithalamin presented several challenges, including the complexity of its extraction process and its stability, leading Khavinson and his team to search for a more practical alternative.

Their efforts led to the isolation and synthesis of Epitalon in the late 1980s. This synthetic tetrapeptide mirrored the beneficial effects of Epithalamin but was easier to produce and more stable. Epitalon was observed to have a profound impact on telomeres — the protective ends of chromosomes mentioned above that shorten over time — potentially slowing down aging at the cellular level.

In the decades following the discovery of Epitalon, Khavinson continued his research, establishing the St. Petersburg Institute of Bioregulation and Gerontology in 1992. The institute has been at the forefront of peptide research, furthering our understanding of their role in biological regulation and aging.

Moreover, Khavinson's work has had a significant impact on the field of biogerontology, contributing to the development of peptide bioregulators, a new class of drugs aimed at slowing down aging and extending a healthy lifespan. His research on Epitalon has been instrumental in demonstrating the potential of peptides as therapeutic agents, paving the way for further developments in this exciting area of research.

The Importance of Epitalon in Anti-Aging Studies

The importance of Epitalon in anti-aging research is rooted in its unique biological role and potential benefits. As mentioned above, Epitalon's significance lies primarily in its hypothesized ability to stimulate telomerase, an enzyme that plays a crucial role in maintaining the integrity of our DNA.

Telomeres

Telomeres are repeating sequences of non-coding DNA that cap the ends of our chromosomes, akin to the plastic tips, called aglets, on shoelaces. They serve a vital function—protecting our genetic data by preventing chromosomes from fraying or sticking to each other, which could lead to genetic scramble or cell death. However, each time a cell divides, the telomeres shorten. This process is natural and inevitable, but it's also directly linked with cellular aging and senescence. As the telomeres shorten, cells progressively lose their capacity to divide and function effectively. This leads to aging-related changes at a cellular level, eventually culminating in cell death.

Telomerase

Often referred to as the "enzyme of immortality," telomerase is a ribonucleoprotein that attaches a sequence of telomere repeats, which varies depending on the species, to the end of telomeres. In essence, it rebuilds and extends telomeres, counteracting their progressive shortening.

However, in most human cells, telomerase activity is low to absent, leading to progressive shortening of telomeres with each cell division. Ultimately, it can contribute to the physical and functional decline associated with aging and in this context, the role of Epitalon becomes critical. It's believed to stimulate the production of telomerase, thereby potentially extending the length of telomeres. By doing so, Epitalon might slow down or even reverse the telomere shortening process, thereby delaying cellular aging and promoting cellular health and longevity.

The Benefits of Epitalon in Anti-Aging

Indeed, Epitalon has garnered substantial interest in anti-aging research due to its unique biological role and potential benefits. The following are some of the reported effects and benefits of Epitalon:

Increased Lifespan:

Epitalon's potential impact on lifespan is one of its most intriguing aspects. This synthetic peptide has been associated with increased lifespan in numerous animal studies, including those on fruit flies and mice. The proposed mechanism behind this effect is Epitalon's ability to stimulate the production of telomerase.

Improved Skin Health:

Epitalon's impact on skin health can be traced back to its role in promoting fibroblast growth. Fibroblasts are cells responsible for producing collagen and elastin, the proteins that give skin its firmness and elasticity. With age, the function and proliferation of fibroblasts decline, leading to reduced collagen and elastin production, resulting in wrinkles and sagging skin.

Epitalon is believed to stimulate fibroblast proliferation and potentially enhance their function, thereby promoting collagen and elastin production and contributing to healthier, more youthful-looking skin.

Enhanced Stress Resistance:

Stress, particularly chronic stress, can have detrimental effects on both physical and mental health and can accelerate the aging process.

Epitalon has shown promise in enhancing resistance to emotional stress. While the exact mechanism remains unclear, it may involve modulation of the hypothalamic-pituitary-adrenal (HPA) axis, the body's central stress response system. By potentially helping to regulate the HPA axis, Epitalon could help improve our physiological response to stress, thereby mitigating its negative effects on health and aging.

Neuroprotective Effects:

Epitalon's potential neuroprotective effects are another area of interest. Aging is often accompanied by cognitive decline and an increased risk of neurodegenerative diseases. Epitalon may exert its neuroprotective effects through several mechanisms, including the reduction of oxidative stress in neuronal cells, modulation of neurotransmitter release, and potentially the stimulation of neurogenesis (the growth of new neurons).

These effects could help protect neurons from age-related damage, potentially improving cognitive function and reducing the risk of neurodegenerative diseases.

Immune Modulation:

Immunosenescence, or the decline in immune function with age, is a key aspect of the aging process. Preliminary studies suggest that Epitalon might help modulate immune function, potentially counteracting some aspects of immunosenescence.

This could involve various mechanisms, such as promoting the proliferation of lymphocytes, enhancing the function of natural killer cells, and modulating cytokine production. By helping to maintain a robust immune response, Epitalon could potentially improve resilience to infections and other diseases in older individuals.

Antioxidant Activity:

Oxidative stress, resulting from an imbalance between free radicals and antioxidants, contributes significantly to aging and age-related diseases. Epitalon is believed to have antioxidant properties, which can help neutralize free radicals and reduce oxidative stress. The potential mechanisms for this effect include the upregulation of antioxidant enzymes, such as superoxide dismutase and catalase, and the downregulation of pro-oxidant enzymes.

By mitigating oxidative stress, Epitalon could contribute to cellular health and longevity, further underscoring its potential in anti-aging interventions.

Interaction with Age-Related Processes in the Body

Epitalon's interaction with age-related processes in the body is multifaceted and complex, involving a diverse range of physiological systems and mechanisms. As such, it influences the aging process and potentially even extends longevity.

Interaction with the Pineal Gland

One of the primary ways that Epitalon interacts with age-related processes is through its impact on the pineal gland. The pineal gland, often referred to as the "third eye," is a small endocrine gland located in the brain. It's responsible for the production and secretion of melatonin, a hormone that regulates sleep-wake cycles and plays a significant role in several biological functions, including antioxidant activity and immune response.

As we age, the function of the pineal gland declines, resulting in diminished melatonin production. This reduction is believed to contribute to several aging-related conditions, including sleep disorders, immune system decline, and increased oxidative stress. Epitalon has been shown to restore melatonin secretion by the pineal gland to levels typically seen in youth, potentially mitigating these age-related changes.

Modulation of the Hypothalamus and the Pituitary Gland

Epitalon also appears to have a modulating effect on the hypothalamus and the pituitary gland, two critical structures in the body's endocrine system. The hypothalamus acts as the body's primary regulator of homeostasis, controlling functions such as temperature, hunger, and thirst. It also influences various aspects of behavior and emotional response. The pituitary gland, often called the "master gland," secretes hormones that regulate many other endocrine glands in the body.

Aging can disrupt the normal functioning of the hypothalamus and pituitary gland, leading to imbalances in hormone levels and disruptions in homeostasis. Epitalon's potential modulation of these structures could help preserve their function, maintaining hormonal balance and homeostasis.

Regulation of Gene Expression

At a more fundamental level, Epitalon may also influence age-related processes by regulating gene expression. Our genes are not static; their activity can be upregulated or downregulated in response to various factors. Epitalon has been shown to alter the expression of several genes associated with aging, including those involved in cell cycle regulation, apoptosis (programmed cell death), and protein synthesis. By influencing gene expression, Epitalon could potentially affect a wide range of biological processes related to aging.

Mitochondrial Function

Lastly, Epitalon may also interact with age-related processes at the level of the mitochondria. Mitochondria, often referred to as the "powerhouses of the cell," generate the majority of our cells' energy supply. However, mitochondrial function declines with age, contributing to reduced cellular energy production and increased oxidative stress.

Epitalon has been shown to enhance mitochondrial function, potentially increasing energy production and reducing oxidative stress. This effect could have broad implications for cellular health and longevity, as well as for the function of tissues and organs throughout the body.

The Impact of Epitalon on Aging Markers in Cells

Epitalon's impact on aging markers in cells has been the subject of extensive research, revealing a variety of intriguing effects. The studies cited above provide a wealth of information on Epitalon's potential role in combating cellular aging. As mentioned earlier, one of the most significant findings is the effect of Epitalon on telomerase activation in human somatic cells. Telomerase is an enzyme that adds DNA sequence repeats to the ends of DNA strands in the telomere regions, effectively reversing the process of telomere shortening that occurs with each round of cell division. This process is crucial for maintaining the integrity of our DNA and preserving cellular health.

Further research has demonstrated Epitalon's potential role in preventing post-ovulatory aging-related damage in mouse oocytes (immature ova), as detailed in a study published in the National Center for Biotechnology Information (NCBI). By preventing early cellular aging processes, Epitalon could potentially preserve the quality and function of oocytes, which typically decline with age.

Epitalon also appears to have a significant impact on hormonal functions, particularly those related to the pineal gland and the pancreas. According to a study published in ScienceDirect, Epitalon could restore age-related disturbances in these hormonal functions. This includes improving the responsiveness of islet β-cells and peripheral tissues to glucose and insulin, which could have implications for age-related metabolic disorders like type 2 diabetes.

Epitalon's influence on thymus function during aging was investigated in a study published in Springer. The research found that Epitalon administration resulted in a down-regulation effect on the CD54, CD69, and HLA DR activation markers of thymic epithelial cells. This suggests that Epitalon could potentially modulate immune function, an aspect that typically declines with age.

Moreover, a study published in MDPI revealed Epitalon's potential role in neurogenesis. The research showed upregulation of all the studied markers: Nestin, GAP43, and β3-tubulin, which are involved in neuronal development and maturation. This suggests that Epitalon could potentially contribute to the regeneration and functional improvement of the nervous system during aging.

Epitalon and Age-Related Hormonal Changes

As individuals progress through life's stages, the human body inevitably undergoes a series of transformations. One of the most profound changes is observed in the endocrine system, the body's complex network responsible for hormone production and regulation. Hormones, the critical chemical messengers, play a pivotal role in numerous physiological processes, including growth, metabolism, mood stability, and reproductive health. However, with advancing age, these hormonal levels fluctuate, exerting a profound impact on various aspects of health and wellness. Such alterations can lead to shifts in energy levels, changes in body composition, variations in cognitive function, and modifications in sexual well-being.

In relation, the mechanisms through which Epitalon influences age-related hormonal changes are complex and multifaceted, involving a variety of biological processes and pathways.

Melatonin Secretion:

Epitalon's impact on melatonin secretion is thought to occur via its interaction with the pineal gland. More specifically, Epitalon, a synthetic version of the pineal peptide, Epithalamin, may mimic the action of natural peptides in the body, stimulating the pineal gland to increase the production and release of melatonin. This not only helps regulate sleep-wake cycles but also has antioxidant properties that protect cells from damage.

Reproductive Hormones:

In regards to reproductive hormones, Epitalon is believed to interact with the hypothalamic-pituitary-gonadal (HPG) axis, a major neuroendocrine system that controls reproductive processes. By modulating the communication within this axis, Epitalon might influence the synthesis and release of sex hormones such as estrogen and progesterone, potentially mitigating the decline in reproductive function associated with aging.

Cortisol Regulation:

Epitalon's effect on cortisol regulation is likely due to its influence on the hypothalamic-pituitary-adrenal (HPA) axis, another critical neuroendocrine system. The HPA axis regulates the body's response to stress, including the production and release of cortisol. Epitalon may help restore the normal rhythm of cortisol production, potentially reducing excessive stress responses and promoting better stress resilience.

Thyroid Hormone Modulation:

As for thyroid function, Epitalon is suggested to interact with the hypothalamic-pituitary-thyroid (HPT) axis. This axis controls the synthesis and release of thyroid hormones, which regulate metabolism and energy production among other functions. Epitalon might modulate the feedback mechanisms within this axis, potentially smoothing out seasonal variations and delaying age-related changes in hormone levels.

Future Research on Epitalon

The future research prospects for Epitalon are extremely promising and could pave the way for significant advancements in our understanding of aging and the management of age-related conditions.

Epitalon's potential impact on genetic mechanisms of aging is one of the most exciting areas of research. Preliminary studies have shown that this peptide can inhibit certain age-related changes in reproductive function in animal models, hinting at possible applications in humans.

This research could lead to breakthroughs in how we treat age-related decline in reproductive health and potentially extend fertility windows. Moreover, scientists are delving into molecular dynamics simulations to study how Epitalon peptides interact with lysine dendrimers. The goal is to see if these dendrimers can form a stable complex with Epitalon, which could improve the peptide's stability, increase its bioavailability, and enhance its therapeutic efficacy. If successful, this research could revolutionize how we deliver and administer Epitalon.

Research is also focusing on the neuroendocrine theory of aging, with studies suggesting that Epitalon's primary mechanism involves its effects on the neuroendocrine system. This system regulates the release of hormones in the body, which play crucial roles in various biological processes, including growth, metabolism, and stress response. Understanding how Epitalon interacts with this system could provide new insights into how we might slow or reverse age-related changes in hormonal function.

Furthermore, Epitalon could play a significant role in cancer research and treatment strategies. Given its ability to stimulate telomerase activity, it may hold promise for the treatment of cancers characterized by abnormal telomere and telomerase dynamics. Epitalon may also play a role in regulating the body's circadian rhythms. Circadian rhythms are physical, mental, and behavioral changes that follow a daily cycle. They affect sleep, feeding patterns, hormone production, and cell regeneration, among other things. By influencing these rhythms, Epitalon could potentially help manage sleep disorders, enhance daytime energy, and improve overall health and well-being.

Another potential mechanism of Epitalon is its impact on the immune system. Some researchers believe that it may help boost the immune response, which naturally diminishes as we age. This enhancement could potentially lead to better defense against common illnesses and diseases, as well as speed up recovery times.

Epitalon's effects on cardiovascular health are also being explored. The peptide may help protect against heart disease by reducing lipid accumulation in the arteries, a key contributor to heart disease. It might also help regulate blood pressure and reduce the risk of stroke. Research also suggests that Epitalon may have neuroprotective properties. It could potentially help protect neurons from damage, reduce the risk of neurodegenerative disorders, and improve cognitive function.

Lastly, researchers are investigating Epitalon's pharmacological properties, particularly its ability to activate the enzyme telomerase in the body. This research has led to the introduction of Epitalon into clinical practice and future studies will likely continue to explore this mechanism.

Documented Effects of Epitalon on Longevity

Research has demonstrated that the administration of this extraordinary peptide, Epitalon, has led to a substantial enhancement in the lifespan of several organisms, including fruit flies and mice. These findings present strong evidence for the potential of Epitalon in life extension, underlining its promising prospects in the realm of anti-aging research and other fields.

Epitalon was evaluated for its potential impact on aging and longevity in a study conducted by Vladimir N Anisimov and colleagues. The research found noteworthy effects of Epitalon on slowing down age-related changes, particularly in reproductive health, where it delayed the cessation of estrous function.

It also contributed to genetic stability, decreasing chromosome aberrations in bone marrow cells by 17.1%, which suggests a role in preventing certain age-related diseases.

The study further identified Epitalon's positive influence on lifespan extension. It increased the lifespan of the last 10% of survivors by 13.3% and the maximum lifespan by 12.3% compared to the control group. 86] Furthermore, while it did not affect the total spontaneous tumor incidence, Epitalon significantly inhibited the development of leukemia six-fold compared to the control group. These findings led researchers to suggest that Epitalon might act as a geroprotector. They concluded that its long-term administration appeared safe in mice, though these results are based on animal models and more human studies are needed. 1]

Other findings have also highlighted the potential of Epitalon in increasing the lifespan of Drosophila melanogaster, a species of fruit fly often used in biological research. According to studies, Epitalon significantly increased the lifespan of imagoes (adult flies) by 11-16% when applied at exceptionally low concentrations during their developmental stage.

The effective concentrations of Epitalon were found to be significantly lower than those of melatonin, a potent antioxidant known for its longevity benefits. 2]

Epitalon and the Maintenance of Healthy Function in Aging Tissues

Due to its unique ability to regulate cell cycles through the upregulation of telomerase activity, Epitalon plays a critical role in the preservation and rejuvenation of aging tissues.

Its potential extends to various aspects of human health. Aside from promoting telomerase activity, Epitalon could also potentially extend the lifespan of cells and maintain the vitality of various tissues in the body by affecting the cardiac tissues. Age-related cardiac dysfunction is a significant concern, often leading to conditions like heart failure.

In animal models, Epitalon has shown promise in preserving cardiac tissue health. It appears to do so by reducing lipid peroxidation, a process that can cause cell damage and by improving myocardial structure and function. These findings suggest that Epitalon could potentially play a role in protecting against age-related heart diseases.

The liver, another organ significantly impacted by aging, also seems to benefit from Epitalon. Studies using aged animal models have shown that Epitalon can enhance protein synthesis in liver cells, a crucial aspect of maintaining liver function. In addition to these benefits, Epitalon has shown promise as a treatment for liver disorders. It is believed to improve wound healing and tissue regeneration not only in the liver but also in other parts of the body such as skin, hair follicles, bones, stomach, and intestinal linings.

In addition to its effects on cardiac and liver tissue, Epitalon may also exhibit neuroprotective properties.

Aging is a significant risk factor for various neurodegenerative disorders, primarily due to the decline in neuronal health and function. Preliminary studies suggest that Epitalon could potentially help protect neurons from age-related damage, thereby preserving cognitive function and reducing the risk of disorders such as Alzheimer's and Parkinson's disease.

Overall, Epitalon stands as a significant development in our perpetual quest for youth and longevity. This peptide represents the convergence of our age-old desire to remain "forever young" through modern scientific advancements. Its potential to extend the lifespan of various organisms is a testament to its promising role in the sphere of anti-aging and longevity research.

CHAPTER 2

THE SCIENCE OF TELOMERES

Our understanding of aging is continually evolving as scientific research advances. Traditionally, aging is associated with physical decline and age-related disability. However, aging does not necessarily have to automatically equate to deterioration. Recent discoveries offer a more optimistic perspective, revealing that our lifestyle choices and the way we live profoundly impact the way we age.

At the center of many of the recent breakthroughs related to the science of aging and longevity are our telomeres. As mentioned above, these telomeres naturally shorten, leading to cellular aging and an increased risk of disease. However, the rate of telomere shortening can be influenced by many of the choices we make in life. Healthier habits, such as maintaining a balanced diet, regular exercise, getting enough sleep, and effectively managing stress can slow down the shortening of our telomeres. On the other hand, detrimental habits like smoking, excessive drinking, and chronic stress can greatly accelerate their shortening and result in what some term "aging before our time."

Preliminary studies have now shown that Epitalon has the potential to extend the lifespan of animals by stimulating an enzyme that preserves telomere length. The implications of this for our understanding of aging and lifespan are significant! These scientific revelations challenge the traditional notion that aging is an inevitable period of decline. We now have the remarkable opportunity to positively influence our aging process and redefine our expectations of what it means to grow older.

Telomeres And Our Cellular Biology

DNA is short for Deoxyribonucleic Acid and is often referred to as "the blueprint of life." It is a complex molecule that carries the genetic instructions which our bodies need to develop, live, and reproduce. This incredible structure is found in every cell of our body and instructs them on how to function, much the same way as a master architect would guide the construction team building a massive and intricate skyscraper. DNA also defines our unique characteristics, from the color of our eyes to our susceptibility to certain diseases or disorders. It is the biological carrier of our genetic legacy, linking us to our ancestors and passing our traits down to our descendants.

Chromosomes are thread-like structures within our cells containing a long string of DNA. DNA could be considered a coded instruction manual with its information organized into specific units, the genes. Therefore, in simple terms, chromosomes are storage units for our DNA which, in turn, contains the genes that determine our physical traits.

Located at the ends of each chromosome are telomeres. These consist of repetitive sequences of non-coding DNA that protect the chromosome from degradation, end-to-end fusion, and improper recombination. Non-coding DNA, also known as "junk DNA," refers to sequences in a genome that do not encode protein sequences. These, despite not producing proteins, play various crucial roles in the function of the body such as regulating gene expression, protecting the integrity of chromosomes, aiding in DNA replication, and contributing to the evolutionary process through genetic variations.

Each chromosome within a cell has two telomeres, one at each end, composed of repeated DNA sequences. In humans, this repeated sequence is TTAGGG, which can be repeated up to 3,000 times.

Along with these DNA sequences, telomeres also contain various proteins, collectively known as The Shelterin Complex. This protein complex helps maintain telomere structure and function.

During DNA replication, an enzyme called DNA polymerase synthesizes a new strand of DNA based on the template strand. However, due to the directionality of DNA synthesis, the very end of the template strand cannot be fully copied.

This phenomenon, known as the "end-replication problem," leads to a progressive shortening of the chromosomes with each cell division. Telomeres serve to compensate for this loss, being the part that's "sacrificed" during replication, thus protecting the crucial genetic information within the chromosomes. Therefore, the function of telomeres is often comically likened to the plastic caps on our shoelaces, preventing them from unraveling.

The Impact of Telomeres on Cellular Aging

The role of telomeres extends beyond merely protecting our genetic material. They have a profound influence on cellular aging, essentially acting as a biological clock within our cells. Each time a cell divides, its telomeres shorten slightly due to the end-replication problem.

When DNA duplicates itself, it does so in a one-way direction, much like how a zipper runs. This process is performed by a worker molecule called DNA polymerase. However, due to this one-directional nature of the replication, DNA polymerase is unable to copy the end of a specific strand, referred to as the "lagging strand." Although it might be a silly analogy, it could be likened to attempting to paint the edge of a wall, yet the handle of your paintbrush is too short to reach the very end. This results in a tiny bit of wall (or DNA) at the end not getting copied each time the cell divides.

In the same manner, telomeres serve as disposable buffers, taking the brunt of this replication limitation and safeguarding the vital genes within the chromosomes from being lost or damaged.

When telomeres reach a critically short length, the cell perceives this as DNA damage.

This perception triggers a cascade of events that culminate in the cell entering a state of growth arrest, or senescence. Senescent cells are also referred to as "zombie cells," as they have ceased to perform normal cellular division and functionality but continue to exist in a state of metabolic activity. In short, they continually use up energy but do not produce anything or perform any function.

These cells have acquired their zombie nickname due to their "undead" status – they are neither truly alive nor completely dead and have the potential to harm neighboring healthy cells. While their primary function was thought to be related to preventing cancerous growth by halting the cell cycle of damaged cells, newer research suggests they may also have roles in tissue repair and wound healing. As senescent cells lose their ability to divide and replenish tissues, this leads to various physical manifestations of aging such as skin wrinkles, muscle weakness, poor memory, and increased susceptibility to diseases.

The importance of telomere length control becomes even more evident when we consider what would happen if a cell bypasses senescence and continues to divide despite having critically short telomeres. This situation pushes the cell into a state of genomic instability, often referred to as a crisis. In this state, the unprotected chromosome ends can fuse, leading to complex rearrangements during cell division. This genomic instability can result in improper DNA repair and mutations, resulting in an elevated risk of diseases like cancer.

However, not all cells are destined to succumb to the effects of telomere shortening. Certain cells, such as stem cells and germ cells, express an enzyme called telomerase.

This remarkable enzyme adds repetitive telomere sequences "back onto" the ends of chromosomes, effectively counteracting the telomere shortening that happens during DNA replication!

This ability allows these cells to maintain their telomere length and extend their lifespan, which contributes to tissue renewal and regeneration. However, most human somatic (body) cells express low levels of telomerase - or none at all - leading to a net loss of telomere length over time. This progressive telomere shortening acts as a molecular clock that ultimately limits the number of times a cell can divide, known as the Hayflick limit. Once this limit is reached, the cell enters a state of permanent growth arrest, causing cellular aging and, by extension, organismal aging.

Assessing Telomere Length and Biological Age

As mentioned earlier, the aging process is characterized by the progressive shortening of telomeres with each cell division. When telomeres reach a critically short length, cells enter a state of growth arrest known as senescence, effectively ceasing to divide or replenish tissues. Assessing telomere length can therefore provide significant insights into a person's biological age. Over the years, several techniques have been developed that allow a deeper understanding of the biology of telomeres.

Quantitative Polymerase Chain Reaction:

One commonly used method is quantitative polymerase chain reaction (qPCR), which relies on detecting and amplifying telomeric DNA sequences by the use of specific primers. By comparing the telomere signal to a reference gene, researchers can estimate the relative length of telomeres in a sample. qPCR is a highly efficient method, allowing for quick results and requiring only small amounts of starting material. This makes it suitable for large-scale studies where comprehensive data can be gathered.

Southern Blot Analysis of Terminal Restriction Fragments:

Southern blot analysis of terminal restriction fragments (TRFs) is another established method. It involves digesting genomic DNA with a restriction enzyme that "cuts outside" the telomeric repeats. The resulting fragments are then separated by gel electrophoresis and transferred with a telomeric probe onto a membrane for hybridization.

By analyzing the distribution of these fragments, we can estimate telomere length.

While this technique may be somewhat labor-intensive and time-consuming, it provides an accurate measurement and allows for the assessment of telomere length heterogeneity.

Molecular Combing:

Molecular combing is a newer technique that offers a more direct measurement of telomere length. It involves stretching individual DNA molecules on a surface, allowing us to visualize and precisely measure telomere length at the single-molecule level. Molecular combing provides information about the absolute length of telomeres and has the potential to detect variations in telomere length within cells. This cutting-edge approach enhances our understanding of telomere dynamics and their impact on cellular health.

These methods, among others, offer valuable insights into the biology of telomeres and their relation to cellular aging. Telomere length serves as a biomarker for cellular aging, reflecting the cumulative effects of DNA replication and damage over time.

While chronological age refers to the number of years since birth, biological age refers to the physiological condition of cells and tissues, which can differ from chronological age.

The Function of Telomerase in Telomere Maintenance

Due to the limitations in the function of DNA replication, telomeres naturally shorten with each cell division cycle. If this shortening process progresses unchecked, it can lead to genomic instability, cellular senescence, or apoptosis — all of which are associated with aging and various diseases. Herein lies the critical role of telomerase.

As an example, genomic instability, which refers to a high frequency of mutations within the genome, can lead to the development of cancerous cells. Cellular senescence, on the other hand, is associated with many age-related pathologies, such as Alzheimer's.

When apoptosis, or programmed cell death, becomes dysregulated, it can contribute to neurodegenerative disorders and cardiovascular diseases. These processes all contribute to our bodies' health and maintenance when functioning properly but drive the progression of various age-related diseases when compromised.

The enzyme telomerase is a ribonucleoprotein enzyme complex that adds specific DNA sequence repeats (the "TTAGGG" found in all vertebrates) to the 3' end of DNA strands in the telomere regions, which are found at the ends of chromosomes. This addition of repeated DNA sequences replaces the sequence loss that occurs during each cycle of DNA replication, thereby preventing chromosomal fraying, preserving genomic integrity, and promoting cellular division and survival.

The telomerase enzyme complex consists of two main components:

1. TERT (telomerase reverse transcriptase), the catalytic protein component, and
2. TERC (telomerase RNA component), an RNA molecule that serves as a template for the synthesis of telomeric repeats.

Telomerase activity is very tightly regulated. In humans and other mammals, telomerase is quite active in embryonic and neonatal development stages but is "silenced" in most of our cells when we reach adulthood.

This silencing leads to progressively further telomere shortening with each cell division and eventually leads to cellular senescence or apoptosis when the telomeres become critically short. However, certain cells that need to divide frequently or indefinitely, such as stem cells, immune cells, or cancer cells, have upregulated telomerase activity. This enables these cells to actually maintain their telomere length and escape senescence or apoptosis.

You could say that these cells are able to "live forever."

In addition to its role in maintaining telomere length, telomerase is involved in various other cellular processes. Emerging research suggests that telomerase may have functions in DNA repair, the regulation of our gene expression, and the modulation of cellular senescence pathways. Understanding the regulation of telomerase activity holds significant implications for aging and disease. In recent years, scientific research has focused on exploring the potential therapeutic applications of telomerase and the activation of telomerase has been proposed as a strategy to slow down the aging process itself!

Telomere Shortening and Age-Related Diseases

The progressive shortening of telomeres is a normal part of aging. However, when telomeres become too short, they can no longer fulfill their protective role. Consequently, the genetic material becomes exposed and vulnerable to damage, which leads to cell malfunction, senescence, or cellular death. This phenomenon of telomere shortening is associated with numerous age-related diseases!

From cardiovascular disorders and neurodegenerative diseases to various types of cancer, the impact of telomere wearing away appears to be far-reaching in relation to human health.

As mentioned, a significant mechanism through which telomere shortening contributes to age-related diseases is through cellular senescence. When telomeres reach a critically short length, our cells undergo a state of replicative senescence where they become unable to divide further. Senescent cells accumulate in tissues throughout the body as the aging process progresses.

These cells release pro-inflammatory molecules, certain growth factors, and other bioactive molecules, collectively known as the senescence-associated secretory phenotype (SASP). SASP can trigger chronic inflammation, impair tissue function, and even promote the development of age-related diseases.

One condition strongly associated with telomere shortening is cardiovascular disease. Telomere shortening in endothelial cells, which line the blood vessels, has been linked to the development of atherosclerosis, hypertension, and other vascular disorders. Shortened telomeres in endothelial cells impair their ability to repair and regenerate, leading to endothelial dysfunction and promoting the progression of cardiovascular diseases. Telomere shortening has also been implicated in certain types of cancer. While cancer cells typically "hack" the system and are able to initiate telomerase reactivation, which allows them to maintain their telomere length, the prior telomere shortening in non-cancerous cells contributes to tumorigenesis.

Tumorigenesis is the process where normal cells transform into cancer cells. This transformation is a complex process that typically occurs over time and involves a series of genetic changes. This process can be triggered by various factors such as genetic mutations, exposure to harmful substances, or certain viral infections. The end result is abnormal cell growth and proliferation leading to the formation of a tumor.

When telomeres become critically short, cells may activate certain DNA damage response pathways, leading to genomic instability and potential oncogenic mutations. Oncogenic mutations refer to changes in the genetic material of a cell that can lead to the development and progression of cancer. The term "oncogenic" itself derives from "onco-" meaning mass or tumor, and "-genic" meaning producing. Hence, oncogenic signifies the potential to cause or produce tumors.

These mutations usually occur in two types of genes:

- Oncogenes
- Tumor suppressor genes

When functioning normally, these genes respectively promote cell growth and inhibit cell growth to maintain a balanced cell cycle. However, oncogenic mutations disrupt this balance and lead to uncontrolled cell proliferation and, eventually, the formation of tumors. Therefore, this telomere dysfunction-driven genomic instability can promote the development of cancer and influence its progression.

Furthermore, telomere shortening has been associated with neurodegenerative diseases such as Alzheimer's disease and Parkinson's disease.

Shortened telomeres in neuronal cells may impair their ability to replicate and repair DNA damage, leading to neuronal dysfunction and increased susceptibility to neurodegeneration. Telomere shortening also affects the immune system, which plays a critical role in maintaining overall health and defending against infections and diseases. As immune cells undergo replication and division to combat pathogens, their telomeres progressively shorten. In time, this telomere attrition can impair immune cell function, leading to compromised immune responses and increased vulnerability to infections and age-related diseases.

Factors Affecting Telomere Length

On a more positive note, our telomere length is not only determined by the passage of time or the number of cell divisions we are genetically predisposed to. It is influenced by a multitude of other factors, including our lifestyle and environmental exposures.

Dietary Influences

Interestingly, research has suggested that our dietary habits can influence telomere length, potentially impacting our health and longevity. For instance, consuming more seeds, nuts, legumes, seaweeds, and coffee were found to elongate telomeres. Dietary fiber intake also appears to be linked with longer telomeres. Moreover, a study revealed that elevated dietary zinc intake was significantly related to longer telomere length among adults aged 45 years and above. 3]

Unhealthy dietary habits, especially an excessive consumption of processed foods, highly refined sugars, and certain food additives, have been linked to the shortening of our telomeres. Processed foods typically contain high levels of unhealthy fats and sodium, while also being low in essential nutrients, thus promoting oxidative stress and inflammation which can accelerate telomere shortening.

Similarly, diets high in refined sugars lead to increased insulin resistance and oxidative stress. Certain food additives, particularly those found in highly processed foods, have been associated with cellular damage which further contributes to the shortening of telomeres.

Dietary habits can also affect our telomere length. Greater adherence to the Mediterranean diet, which is rich in fruits, vegetables, whole grains, and olive oil, has been associated with longer telomere lengths. This could be due to the diet's high content of antioxidants and anti-inflammatory nutrients, which help protect telomeres from damage.

Genetic Factors

Genetic predisposition also plays a significant role in determining telomere length. Several genes have been identified that play a role in telomere maintenance, primarily through their involvement in telomerase activity. One of the most notable is "TERC." TERC stands for Telomerase RNA Component and is a crucial part of the human genome.

It refers to the RNA component of the telomerase enzyme, which is essential for the replication and protection of chromosomes during cell division. Any mutation in the TERC gene could potentially lead to disorders associated with premature aging and a shortened lifespan.

Variations or mutations in the TERC gene can also lead to an inability to adequately replenish telomeres, resulting in significantly shorter ones. For instance, certain mutations in the TERC gene have been associated with dyskeratosis congenita which is a rare inherited condition characterized by premature aging, bone marrow failure, and increased susceptibility to cancer. Individuals with this condition often have extremely short telomeres due to impaired telomerase activity.

Another gene of interest is "TERT," the catalytic component of telomerase. TERT stands for Telomerase Reverse Transcriptase and it is a protein-coding gene that is involved in the protection of chromosomes from degradation. By adding DNA sequence repeats to the 3' end of DNA strands in the telomere regions, it plays a crucial role in cellular aging and cancer.

Mutations in the TERT gene can also lead to reduced telomerase activity and shorter telomeres. For example, TERT mutations have been linked to idiopathic pulmonary fibrosis which is a disease characterized by the progressive scarring of the lungs, and certain types of cancer. Beyond TERC and TERT, several other genes have been associated with our telomere length. OBFC1 is one such gene, which encodes a protein involved in the regulation of telomerase, known as STN 1. This forms a part of the CST complex (CTC 1-STN 1-TEN 1).

This complex is crucial for telomere replication and capping and plays a significant role in maintaining overall telomere length and stability. Variations in the OBFC1 gene can impact the efficiency of the CST complex and so influence telomere length.

It has been suggested that certain genetic variants of OBFC1 are associated with longer leukocyte telomere length (LTL), which is a commonly used marker of biological aging. Moreover, these variants have also been linked with a lower risk of coronary heart disease, potentially due to their protective effects on telomeres.

CTC1 is another gene that encodes a component of the CST complex. The protein CTC 1 works in conjunction with STN 1 and TEN 1 to protect telomeres from degradation, prevent telomere elongation by telomerase, and aid in efficient telomere replication during the S phase of the cell cycle. The S phase, also known as the Synthesis phase, is a crucial part of the cell division cycle. It is during this phase that the DNA of a cell is replicated or copied. Think of it as creating a duplicate set of blueprints.

This process is essential for the proper growth and development of any organism and for the repair and maintenance of tissues in an adult organism. It's a very complex process but cells have an inbuilt mechanism to correct most mistakes, ensuring the DNA is copied accurately for the next generation of cells.

Mutations in the CTC1 gene can lead to disorders characterized by short telomeres, such as Coats plus syndrome and dyskeratosis congenita. These conditions are marked by a variety of symptoms, including premature aging, increased cancer risk, and various multi-system pathologies.

Lifestyle Factors

Various lifestyle factors can greatly affect our telomere length. One of the most potent lifestyle influences on the length of our telomeres is physical activity. It has been found that individuals who engaged in regular, vigorous exercise had significantly longer telomeres than those who were sedentary.

The mechanisms behind this relationship are multifaceted. Regular physical activity reduces oxidative stress and inflammation, two key processes known to accelerate telomere shortening. Exercise also enhances the activity of telomerase, and thus preserves telomere length. Furthermore, exercise can stimulate the release of growth hormones and other factors that may aid in telomere maintenance. Unhealthy habits like smoking and excessive alcohol consumption have been associated with hastened telomere shortening. Both of these factors can increase oxidative stress and cause direct DNA damage, which in turn can lead to accelerated telomere erosion.

Cigarette smoke contains numerous harmful chemicals that can induce oxidative stress and inflammation, thereby damaging DNA and telomeres. A significant negative correlation has been found between pack-years of smoking and telomere length, suggesting a cumulative effect of smoking on telomere attrition.

Similarly, excessive alcohol consumption can increase oxidative stress and inflammation in the body. It can also lead to folate deficiency, which may further impair DNA repair and synthesis, affecting telomere maintenance. Research has found that heavy alcohol use is associated with shorter telomere length.

Psychological Stress

Chronic psychological stress is another significant factor that can affect telomere length. When we encounter a stressful situation, our body responds by initiating the "fight or flight" response. This process involves the release of various hormones, like cortisol, adrenaline, and noradrenaline, which prepare the body to respond to the perceived threat.

However, when stress becomes chronic, this hormonal response can lead to increased oxidative stress and inflammation. Oxidative stress refers to an imbalance between the production of harmful free radicals and the body's ability to neutralize them. These free radicals can damage various cellular components, including DNA and telomeres. Inflammation, on the other hand, is a part of the body's immune response. However, when it becomes chronic, it can also contribute to cellular damage and has been associated with shorter telomeres.

Cortisol is often referred to as the "stress hormone" and is released in larger amounts during prolonged times of stress. While it plays a crucial role in the stress response, prolonged exposure to cortisol can have detrimental effects on the body. High cortisol levels have been linked to increased oxidative stress and inflammation, which can accelerate telomere shortening. Moreover, cortisol can also reduce the activity of telomerase. Telomerase adds those repetitive DNA sequences to the ends of chromosomes as talked about above, effectively rebuilding and preserving telomeres. However, studies have shown that chronic stress and high cortisol levels can inhibit telomerase activity, leading to faster telomere erosion.

The perception of stress and the ability to cope with it can also influence telomere length. Individuals who perceive more stress or who cope poorly with stress may experience more significant hormonal fluctuations, leading to more oxidative stress, inflammation, and potential telomere shortening.

Environmental Exposures

Environmental exposures refer to the contact we have with various elements in our environment, including air, water, soil, and the different substances they contain. These exposures can significantly impact human health, particularly at the cellular level, where they can influence the length of telomeres.

Air pollution is a significant environmental exposure that has been linked to telomere shortening. Fine particulate matter (PM2.5), a common component of air pollution, can infiltrate the body's cellular system when inhaled. PM2.5 is known to induce oxidative stress and inflammation, two key processes that can accelerate telomere shortening.

Long-term exposure to PM2.5 has been associated with shorter telomere length in adults. This association suggests that air pollution may contribute to cellular aging and related health effects, such as cardiovascular disease and certain types of cancer.

Exposure to certain chemicals, either in the environment or the workplace, can also affect telomere length. For instance, studies have shown that exposure to lead, a toxic heavy metal, is associated with shorter telomere length in children. This finding suggests that early-life exposure to environmental toxins can impact cellular aging processes, potentially increasing disease susceptibility later in life.

Another study found that exposure to polycyclic aromatic hydrocarbons, a group of chemicals formed during the incomplete burning of coal, oil, gas, wood, garbage, and other organic substances, influenced telomere length in preschool children. 4] This association underscores the potential health risks of exposure to environmental toxins at a young age.

Age and Sex

As expected, age is a significant factor influencing telomere length, with telomeres generally shortening as we grow older. However, the rate of this shortening can vary widely among individuals of the same age, reflecting the influence of other factors mentioned above.

This variability reflects how diverse genetic, environmental, and lifestyle factors influence the way we age. For instance, a person who smokes, is physically inactive, and eats a poor diet may have shorter telomeres than a person of the same age who does not smoke, exercises regularly, and eats a healthy diet.

Sex is another factor that can influence telomere length. On average, women tend to have longer telomeres than men. This difference has been observed across various age groups and populations, suggesting a fundamental biological distinction between the sexes. One possible explanation for this difference lies in the hormone estrogen, as estrogen is known to stimulate the production of telomerase. Women may therefore have increased telomerase activity and consequently longer telomeres, as they have higher levels of estrogen than men.

In addition, the second X chromosome in females could possibly offer a certain protective effect. The "extra copy" of certain genes may help maintain the stability of cells, including the maintenance of telomere length. Overall, aging no longer has to result in inevitable deterioration.

Aside from the many powerful regenerative therapies and modalities developed or Epitalon's incredible efficacy, maintaining a balanced lifestyle, adequate rest, proper nutrition, and regular exercise can significantly affect the length and health of our telomeres. The cumulative knowledge we have amassed about these incredible lifestyle factors empowers us to take control of our well-being, thereby extending our vitality and wellness.

CHAPTER 3

EPITALON AND TELOMERE ACTIVATION

The field of research in aging has experienced a major paradigm shift over the past few decades. Traditionally, age-related illness and deterioration have been viewed as a natural part of the later phases of life that we all must face. However, the traditional concept of aging may not be as "natural" or "inevitable" as we may believe. Central to a newer understanding of aging is the idea that aging is not just a chronological process but a biological process, closely tied to the state of the cellular health of our bodies. At the heart of this process are the telomeres, covered extensively in the earlier chapter. As mentioned in Chapter two, telomere length and the activity of the enzyme telomerase have been found to play a pivotal role in cellular aging and, by extension, the way our bodies age.

However, nature itself has devised an inbuilt "key to eternal life" that can counteract the shortening of telomeres: the enzyme telomerase used to manage this process!

Telomerase and Chromosomal Integrity

Telomerase is a unique and specialized ribonucleoprotein enzyme complex with an essential role in cellular biology. It is primarily tasked with the maintenance of our genomic integrity by the extension of telomere lengths. As mentioned, the function of telomerase is critical in facilitating cellular division and counteracting cellular aging, which are the two processes centered at the heart of life's continuity and health.

The telomerase enzyme complex, which is responsible for maintaining the telomeres, is composed of two core components: the telomerase reverse transcriptase and the telomerase RNA component, or the TERT and TERC, respectively.

The TERT is a catalytic protein component of the telomerase complex. Its primary role is to add DNA sequence repeats to the telomeric ends of the chromosomes. It achieves this through its reverse transcriptase activity, which allows it to synthesize DNA from an RNA template. On the other hand, TERC serves a dual role within the telomerase complex. Firstly, it acts as the template for the synthesis of telomeric repeats added by the TERT subunit. Secondly, it contributes to the proper folding and assembly of the telomerase complex, ensuring its stability and functionality.

One vivid example of actual cellular immortality due to telomerase activity can be seen in human embryonic stem cells. These cells, sourced from the inner cell mass of the blastocyst (an early-stage preimplantation embryo), are characterized by their ability to differentiate into any cell type in the body. They are also renowned for their capacity to self-renew or proliferate without differentiation. The H9 line of human embryonic stem cells, which was derived by Thomson et al. in 1998, illustrates this perfectly.

Despite being over two decades old, these cells continue to grow and divide in laboratories across the globe! This longevity is attributed to the active telomerase they express, which continually elongates the telomeres at the ends of their chromosomes. This prevents the telomeric shortening that would otherwise occur with each cell division and effectively bypasses the "Hayflick limit," which bestows these cells with a form of immortality.

Epitalon and Telomerase: Triggering Rejuvenation

Epitalon (also sometimes known as Epithalone) is a synthetic tetrapeptide originally developed by the Russian scientist Vladimir Khavinson. Dr. Khavinson based his design on a naturally occurring peptide known as Epithalamin (sometimes spelled Epithalamine) secreted by the pineal gland. This peptide has been associated with regulating the body's internal clock through its influence on our circadian rhythm.

Understanding Epitalon's Role in Cellular Aging

Epitalon's renowned anti-aging effects are primarily linked to its capacity to stimulate the production of telomerase. It is believed to involve the binding to specific receptors on the cell surface, which triggers a cascade of biochemical events leading to the activation of the TERT gene, which codes for the catalytic subunit of telomerase. By promoting telomerase production, Epitalon can help maintain telomere length, enabling cells to divide more times before entering senescence.

This could potentially slow down the aging process at the cellular level, leading to improved cell health and longevity.

Epitalon not only stimulates the production of this enzyme; it also appears to enhance its activity. This enhancement may occur through various mechanisms, such as increasing the affinity of telomerase for the telomeric substrate, promoting the assembly of the telomerase complex, or protecting telomerase from inhibitory factors. By boosting telomerase activity, Epitalon could increase the rate at which telomeres are extended, allowing cells to divide more times before reaching senescence. Beyond its influence on telomerase, Epitalon has been associated with the regulation of melatonin production.

Melatonin is also produced by the pineal gland and regulates our sleep-wake cycle or circadian rhythm. Disturbances in this natural sleep-wake cycle have been linked to a variety of health issues, including accelerated aging. Epitalon has been shown to interact with the pineal gland and other components of the endocrine system to regulate the synthesis and release of melatonin, thereby helping to normalize the circadian rhythm. This regulatory effect contributes to the overall anti-aging effects of Epitalon.

Clinical Insights: The Research Behind Epitalon

Current research suggests that Epitalon has a significant impact on telomerase activation and telomere lengthening. For instance, the addition of Epitalon to a culture of telomerase-negative human fetal fibroblasts was found to induce the expression of the catalytic subunit of telomerase, increase the enzymatic activity of telomerase, and lengthen telomeres. [4]

This suggests that Epitalon may be capable of reactivating the actual telomerase gene in somatic cells. This is a significant finding, as it implies that the lifespan of a cell population and potentially even of the whole organism could be extended through the use of this peptide. This research was conducted by Vladimir Khavinson, I. E. Bondarev, and A. A. Butyugov from the St. Petersburg Institute of Bioregulation and Gerontology. Their findings were published in the "Bulletin of Experimental Biology and Medicine" in July 2003. The study supports the potential of Epitalon as an anti-aging intervention and contributes to our understanding of the mechanisms through which this peptide could influence cellular aging.

Furthermore, Epitalon has also been demonstrated to influence the pineal gland and lead to a rejuvenation of melatonin production akin to that seen in our youth. Beyond this, Epitalon has been shown to have multiple corrective effects on endocrine functionality that typically deteriorate when we age.

For instance, when aging rhesus monkeys were administered Epitalon injections, their nocturnal melatonin production saw an increase and the usual daily pattern of cortisol secretion was reestablished, which overall contributed to the normalization of circadian rhythms. 5]

Therapeutic Applications of Epitalon

Epitalon holds incredible benefits! But what conditions could it potentially help alleviate, and what systems or organs of the body may it specifically benefit?

Potential Candidates for Epitalon Therapy:

Given its potential anti-aging properties, Epitalon can be beneficial for those experiencing signs of aging, like decreased energy levels, cognitive decline, or weakened immune function. As Epitalon is believed to slow the cellular aging process, it may help to maintain health and vitality longer. Epitalon has been the subject of extensive scientific research due to its potential therapeutic applications in a variety of health-related fields, such as the following:

Anti-Aging:

Epitalon's anti-aging effects are primarily attributed to its ability to stimulate telomerase activity and lengthen telomeres. This ability to delay cellular senescence could postpone the onset of age-related diseases such as cardiovascular disease, neurodegenerative disorders, and cancer. This is because cellular aging contributes significantly to the development of these conditions.

Beyond this, Epitalon is also believed to play a role in maintaining the structural integrity of cells. It is suggested that by ensuring the health of the cellular structure, tissue function can be improved. This has wide-reaching implications for organ health as well since tissues collectively form organs. By enhancing tissue function,

41

Epitalon could potentially enhance the functionality and health of various organs, contributing to overall vitality.

In essence, the anti-aging properties of Epitalon are tied to its ability to stimulate telomerase, protect telomeres, and maintain cellular structure. These actions at the microscopic level can have macroscopic effects, potentially improving overall health and delaying the onset of age-related diseases.

Neuroprotection:

Research suggests that Epitalon exerts certain neuroprotective effects that may safeguard the brain from damage and the cognitive decline normally associated with aging. These effects might be due to Epitalon's potential to stimulate neuronal differentiation, as well as its potential antioxidative properties.

For instance, epigenetic regulation of gene expression is vital for maintaining cognitive function and neuronal differentiation. This understanding is crucial in studying Alzheimer's disease (AD), commonly linked to DNA methylation, chromatin remodeling, histone modifications, and non-coding RNAs regulation.

AD's pathogenesis is associated with tau protein and amyloid peptides misfolding, mitochondrial dysfunction, oxidative stress, impaired energy metabolism, blood-brain barrier destruction, and neuroinflammation, ultimately leading to damaged synaptic plasticity and memory loss. Ultrashort peptides, with their broad activity range and no reported side effects, are potential neuroprotective agents. In a review by Ilina and colleagues (2022), the epigenetic mechanisms of ultrashort peptides' neuroprotective action in AD were examined.

The role of short peptides in AD pathophysiology is emphasized, hypothesizing that peptide regulation of gene expression might be mediated through interactions with histone proteins, cis and trans regulatory DNA elements, and effector molecules.

The formulation of therapeutic agents based on ultrashort peptides presents a promising addition to AD's multifunctional treatment. 6] In the context of neuroprotection research, Epitalon is suggested to possess certain neuroprotective effects that could potentially protect the brain from damage and age-related cognitive decline, such as in the case of AD.

These effects could be attributed to Epitalon's potential to stimulate neuronal differentiation and its possible antioxidative properties. Epitalon, much like the aforementioned ultrashort peptides, could promote neuronal differentiation, a fundamental process in neurodevelopment and the maintenance of cognitive function.

Firstly, Epitalon may stimulate neuronal differentiation, the process by which cells develop into neurons. This is a key process in neurodevelopment and the maintenance of cognitive function as we age. By promoting the growth and development of new neurons, Epitalon could potentially combat the decline in cognitive function often observed in aging.

Secondly, Epitalon may also exert antioxidative effects. Oxidative stress, resulting from an imbalance between the production of reactive oxygen species and the body's ability to counteract or detoxify their harmful effects, is a major contributor to neuronal damage in neurodegenerative diseases. By bolstering the brain's antioxidant defenses, Epitalon can potentially safeguard neurons from this oxidative damage.

Cancer Prevention:

Epitalon has also displayed potential anticancer properties in animal studies. In these studies, Epitalon was shown to decrease experimental carcinogenesis. Carcinogenesis is a complex process, where healthy, normal cells are transformed into malignant cancer cells.

This transformation is a gradual progression that operates at multiple levels - cellular, genetic, and epigenetic - ultimately leading to the formation of a malignant tumor.

The factors contributing to carcinogenesis are diverse, ranging from genetic predispositions to external elements such as harmful environmental exposures. What is common among these factors is their ability to damage the DNA within cells. In a normal scenario, our cells have mechanisms to repair such damage.

However, with carcinogenesis, these repair mechanisms fail, leading to uncontrolled cell division.

Epitalon's anticancer properties appear to hinge on its influence over the activity of the enzyme telomerase. It is suggested that Epitalon inhibits the rampant cell division characteristic of cancer cells by regulating telomerase activity, thereby limiting the unwarranted elongation of telomeres. This action, in turn, prevents the cells from becoming immortal and uncontrollably proliferative. However, while doing so, Epitalon also ensures the longevity of telomeres in normal cells, thus promoting their health and functioning.

Eye Health:

Epitalon has also been observed to regulate the function of the retina, suggesting potential benefits for eye health. The retina, a thin layer of tissue lining the back of the eye, plays a critical role in vision as it receives light that the lens has focused, converts the light into neural signals, and sends these signals to the brain for visual recognition.

Epitalon's influence on retinal health seems to revolve around its potentially protective and restorative effects on retinal cells. On a molecular level, it is suggested that Epitalon could enhance the functional performance of the retina, possibly by optimizing the cells' metabolic processes, or even by boosting their resilience against stress and damage.

In addition, Epitalon may help to prevent degenerative changes in the retina. This is particularly relevant as the retina, like many other tissues in the body, can be prone to wear-and-tear as we age. Degenerative changes can lead to a decrease in the quality of vision and, in severe cases, vision loss.

Epitalon's impact on these degenerative changes may offer an avenue to help maintain overall retinal health, and by extension, visual health.

Moreover, the potential benefits of Epitalon extend to the prevention of age-related vision loss. One of the most common forms of this is macular degeneration, a condition characterized by the gradual deterioration of the macula (the central part of the retina) leading to central vision loss. Though extensive clinical trials are still required, the potential ability of Epitalon to improve retinal cell function and inhibit degenerative changes could make it a valuable tool in the fight against age-related vision loss.

Immune Function;

Epitalon could also enhance our immune function. A vital component of this immune-enhancing effect is Epitalon's impact on T-cell production. T-cells, otherwise known as T lymphocytes, are a type of white blood cell integral to our immune response. They recognize and destroy infected or malignant cells in the body. Research suggests that Epitalon may stimulate the production of these T-cells, thus enhancing the body's capacity to fight off infections and diseases.

In addition to boosting T-cell production, Epitalon also appears to augment the production of antibodies, another crucial player in our immune defense.

Antibodies are proteins produced by B lymphocytes in response to harmful invaders, such as bacteria and viruses. They latch onto these foreign substances, marking them for destruction by other immune cells. Through up-regulating antibody production, Epitalon could potentially amplify our body's natural defenses.

Beyond its role in T-cell and antibody production, Epitalon seems to regulate cytokine function. Cytokines are a type of protein that play a significant role in cell signaling, affecting the behavior of cells around them. They are pivotal to the immune response, coordinating the body's reaction to infection and inflammation. Epitalon's influence on cytokine regulation could further enhance immune function, facilitating a more effective and coordinated response to threats.

Administering Epitalon

Administering Epitalon effectively and safely is a crucial aspect of leveraging its potential benefits. Fortunately, it can be administered in various ways depending on the individual's comfort level, the dosage required, and the desired effect.

By Injection

The most common and arguably most effective method of administering Epitalon is through subcutaneous injections, which are injections given under the skin. The abdominal region is often chosen as the injection site due to its ease of access. This method directly introduces the peptide into the body, allowing for maximum absorption and utilization.

The dosage of Epitalon can vary, but a common protocol suggests administering 10mg of the peptide daily, either in one single injection or split between two separate injections.

Some sources suggest injecting 20 units subcutaneously daily, with cycles that are repeated 2-3 times per year. These dosages and schedules should be tailored according to individual needs and under the guidance of a healthcare professional.

As Nasal Sprays

Epitalon can also be administered via nasal sprays. This method is very non-invasive and easy to use, which makes it a popular choice for those who are too uncomfortable with injections. While the absorption rate may not be as high as with injections, nasal sprays still provide a direct route into the bloodstream, bypassing the digestive system and preserving the integrity of the peptide.

Through Oral Administration

Oral administration of Epitalon is another option. This method is also very non-invasive and user-friendly. However, it is important to note that peptides, when taken orally, must pass through the digestive system where they may be broken down by the stomach acid before reaching the bloodstream, potentially reducing their effectiveness.

With Topical Applications

Applying Epitalon directly on the skin is another method of administration. This can be particularly useful for targeting specific areas of the body, although the overall systemic absorption may be lower compared to other methods.

Supplementing Epitalon

Epitalon can also be taken as a supplement, such as in the form of Ageless Capsules. Epitalon is a synthetic peptide that is not naturally produced in the human body. However, it is designed to mimic a naturally occurring peptide known as Epithalamin, produced in the pineal gland (a small endocrine organ located near the center of the brain). The goal of these supplements is to boost the body's natural production of Epithalamin, by supplementing it with the synthetic Epitalon.

Monitoring the Effects: How Can Progress Be Tracked?

Monitoring the effects of Epitalon, particularly its impact on telomere length and cellular aging, is a critical aspect of assessing the peptide's efficacy. Healthcare providers employ a combination of methods ranging from direct measurement of telomere length to observing physiological changes indicative of a reversal of signs of aging.

Direct Measurement of Telomere Length:

To assess the impact of Epitalon, healthcare providers typically measure telomere length before and after therapy. This is typically performed using laboratory techniques such as:

Quantitative PCR:

This technique quantifies the amount of DNA in a sample, allowing for a relative measure of telomere length.

Southern Blot of Terminal Restriction Fragments:

This process involves extracting DNA, cutting it into fragments using restriction enzymes. Specific sequences are then identified — in this case, telomeres — through a hybridization probe.

Fluorescence In Situ Hybridization (FISH):

This method involves using fluorescent probes that bind to specific parts of the DNA. This enables visualization and measurement of telomeres under a microscope.

These techniques provide a tangible, quantifiable measure of telomere length, offering a clear assessment of whether Epitalon treatment has led to telomere elongation.

Monitoring Physiological Signs of Aging Reversal

Alongside direct measurements, healthcare providers also monitor physiological changes that indicate a reversal of the common signs of aging. Epitalon is typically associated with a wide range of health benefits, such as improved skin health, better sleep patterns, enhanced immune response, and increased energy levels. Progress in these areas can provide indirect but meaningful signs of the peptide's efficacy.

Aging is a multi-dimensional process that affects our bodies on various levels, from cellular changes to shifts in organ function and overall health. Observable signs of aging can include decreased skin elasticity, changes in sleep patterns, a weaker immune response, and reduced energy levels. On a more systemic level, traditional aging can lead to cognitive decline, cardiovascular issues, and altered metabolic functions.

Regular assessments are essential to track and monitor the state of these changes. A healthcare provider can monitor organ function tests, cognitive function, cardiovascular health, and metabolic parameters to gauge the effectiveness of Epitalon.

As an example, blood tests can reveal changes in cholesterol levels or glucose metabolism, while cognitive assessments can detect improvements in memory or executive function. Cardiovascular health can be assessed through blood pressure measurements, electrocardiograms, or stress tests.

Studies suggest that Epitalon has certain protective effects on various tissues and organs, helping to maintain their function despite the progression of aging.

This means that regular monitoring of organ functions, such as liver enzyme levels, kidney function tests, or lung function tests, can provide valuable insights into the state of overall organ function and in turn the possible efficacy of Epitalon.

Cellular and Molecular Markers Assessment

The assessment of cellular and molecular markers is a crucial part of evaluating the effects of Epitalon. This process involves examining the expression of specific proteins and other biomarkers associated with aging and cellular health. Cellular aging is a very complex process defined by various changes at a molecular level. This include alterations in protein synthesis, DNA damage, and changes in our gene expression, amongst others. Specific proteins or sets of proteins, often referred to as "biomarkers," can therefore serve as indicators of these changes.

One such biomarker is Caspase-3, which is a protein that is involved in apoptosis or programmed cell death. Apoptosis is a natural part of the cellular lifecycle, but when it occurs prematurely or excessively it can contribute to premature aging and various age-related diseases. Caspase-3 is a critical executioner of apoptosis, helping to dismantle the cell once the apoptotic process has been initiated.

Aside from increased telomerase activity, Epitalon has also been shown to inhibit the synthesis of Caspase-3. This could potentially slow down the rate of apoptosis, allowing cells to live longer and thus contributing to an overall anti-aging effect. Therefore, a reduction in Caspase-3 levels following Epitalon treatment could indicate its efficacy. Other molecular markers can also provide insight into the effects of Epitalon on cellular health. These may include markers of oxidative stress, inflammation, DNA damage, and more.

For instance, Epitalon has been shown to decrease the expression of p16INK4a, which is a marker of cellular senescence, and to reduce levels of reactive oxygen species (ROS), which are associated with oxidative stress and cellular damage.

Looking Ahead: The Future of Epitalon and Telomere Research

The future of Epitalon and telomere research is an exciting realm to explore. As scientists continue to delve deeper into the intricacies of cellular aging, the role of Epitalon and its impact on telomeres is becoming increasingly significant. The ongoing and future research holds immense potential to unlock new therapeutic applications and strategies to combat age-related diseases and possibly extend the human lifespan!

As we look ahead, one of the critical areas of focus will be to understand the precise mechanisms through which Epitalon exerts its biophysical effects. While it's known that the peptide stimulates telomerase activity, the exact pathways involved in this process are not entirely understood. Uncovering these mechanisms will provide valuable insights into how Epitalon functions and how it can be used most effectively.

Moreover, future research will also aim to further determine the optimal dosing and administration schedules for Epitalon. This will involve conducting extensive clinical trials to evaluate the peptide's safety, efficacy, and potential side effects in larger populations over extended periods.

Advancements in Telomere Research

On the other hand, telomere research is poised to make significant strides in the coming years. Scientists increasingly recognize the role of telomeres in aging, and a wide range of age-related diseases, such as cancer, cardiovascular disease, and neurodegenerative disorders. Emerging research is currently exploring the possibility of using telomere length as a biomarker for these diseases, which could transform diagnostic procedures and enable earlier and more accurate detection.

Moreover, researchers are also investigating strategies to protect and repair telomeres, which could lead to novel treatments for age-related diseases.

Looking further to the future, Epitalon can revolutionize the field of anti-aging as we know it! This peptide could potentially be used as a therapeutic agent to entirely prevent, let alone delay, the onset of the most common age-related diseases and improve quality of life, while extending the human lifespan to new and yet unknown limits.

As we continue to advance our understanding of cellular aging and the role of telomeres, new therapeutic strategies and applications will transform our approach to aging and age-related diseases!

CHAPTER 4

CLINICAL STUDIES ON EPITALON

Longevity and anti-aging research have been a cornerstone of medical exploration for centuries. The quest to extend the human lifespan, maintain health and vitality, and stave off the inevitable effects of aging has led scientists down many interesting avenues of study. One such field that has captured significant attention is peptides, due to their therapeutic effects and versatility in application. Researchers have now discovered that certain peptides can influence our body's aging processes, potentially extending our health span and delaying the onset of age-related diseases.

Current research in longevity and anti-aging seems to push the boundaries of what we previously thought was humanly possible. Through clinical research, reviews, and research studies, we continually surprise ourselves and discover new data on the power of peptides and novel ways to promote true health and prolong vitality. A particularly intriguing area of peptide research revolves around the peptide Epitalon. This tetrapeptide has been associated with many benefits in relation to our quality of life, increased lifespan, and anti-aging.

Epitalon and Anti-Aging Research

The interest in Epitalon mainly stems from its proven potential to slow the aging process. Aging is a complex process, characterized by a gradual decline in physiological functions and an increased risk of what we term "age-related diseases," such as cancer, cardiovascular disease, and neurodegenerative disorders.

Major Clinical Trials Conducted on Epitalon

Anticarcinogenic Potential

One of the most significant clinical trials conducted on Epitalon was a unique 15-year study focusing on the anticarcinogenic effects of peptides. Anticarcinogenic refers to any substance or process that works to prevent, inhibit, or halt the development of cancer. Carried out at the NN Petrov Research Institute of Oncology, this research aimed to explore Epitalon's potential in inhibiting spontaneous tumor development. This was achieved by administering Epitalon to experimental subjects and closely monitoring them for signs of tumor development throughout the study. The researchers also set up controls for other factors that could potentially influence tumor growth, such as the subjects' overall health, genetics, and environmental influences, to ensure that any observed effects could be attributed to Epitalon.

Researchers found that Epitalon administration resulted in a significant decrease in spontaneous tumor development compared to a control group. This suggests that Epitalon could potentially be used as a novel strategy for cancer prevention.

In addition to these findings, the researchers also reported some interesting observations regarding the anti-aging effects of Epitalon. For instance, they noted improvements in various physiological parameters commonly associated with aging, such as decreases in inflammation and oxidative stress levels. These changes suggest that Epitalon may have broad anti-aging effects beyond its anticarcinogenic properties. 7]

Neurogenesis

Another notable study investigated Epitalon's influence on neurogenesis, specifically its effect on gene expression and protein synthesis. Neurogenesis, or the formation of new neurons within the brain, is vital for maintaining cognitive function and overall brain health. With aging, the ability of the brain to generate new neurons diminishes, leading to cognitive decline and other neurological issues.

In this research, the scientists analyzed the effect of Epitalon on gene expression and protein synthesis, two fundamental biological processes that are involved in neurogenesis. Gene expression refers to the process by which information from a gene is used to create a functional product, usually a protein, while protein synthesis is the process by which cells build proteins.

To measure these effects, the researchers administered Epitalon to experimental subjects and then used techniques such as quantitative PCR (polymerase chain reaction) and Western blotting. Quantitative PCR is a technique used to measure the amount of specific RNA (or gene expression), while Western blotting is used to detect specific proteins in a sample.

The results of the study indicated that Epitalon significantly increased the expression of certain genes involved in neurogenesis. This suggests that Epitalon could potentially enhance the brain's capacity to generate new neurons, thereby slowing down age-related cognitive decline. Moreover, Epitalon also influenced protein synthesis in a way that favored neurogenesis. Proteins are the "workhorses" of the cell and carry out most of the vital functions necessary to sustain life. Therefore, an increase in protein synthesis could mean improved cellular function and, in this case, enhanced neurogenesis. These findings provide compelling evidence of the potential anti-aging effects of Epitalon, particularly concerning cognitive health. 8]

Biomarkers Indicative of Cellular Aging

Epitalon's effects on biomarkers indicative of cellular aging have been extensively studied. In one study conducted on female Swiss-derived SHR mice, researchers found that Epitalon improved several key biomarkers of aging, suggesting that it could potentially slow down the aging process at a cellular level. From the time they were 3 months old until they passed away naturally, a particular breed of female mice were given small injections for five consecutive days every month.

These injections contained either a harmless saltwater solution (used as a control) or a tiny amount of Epitalon. Each group contained 54 mice. The results showed that Epitalon slowed the decline of reproductive functions with age and reduced the likelihood of changes in bone marrow cells that can lead to disease. Additionally, it prolonged the life of the longest-living 10% of mice by about 13.3% and increased the maximum lifespan by 12.3% when compared to the control group. The study noted that while Epitalon, in this study, did not impact the overall occurrence of spontaneous tumors, it significantly reduced the likelihood of developing leukemia when compared with the control group. 13]

This study, published in the International Journal of Cancer, is one of several that demonstrate the promising potential of Epitalon as an anti-aging agent. However, while these results are significant, it is vital to point out that further research is required to fully understand the mechanisms at play and the potential implications for human health. 9]

Specific Anti-Aging Benefits Observed in Clinical Studies of Epitalon

Cellular Regeneration

Cellular regeneration is an essential process in maintaining skin health. It involves the replacement of old, damaged cells with new, healthy ones. This ongoing process slows down as we age, contributing to visible signs of skin aging.

However, Epitalon has been noted for its potential to enhance cellular regeneration. Studies have also suggested that Epitalon, as it is a short form of a peptide consisting only of four amino acids, has the ability to penetrate the cell. 10] This deep level of penetration can allow Epitalon to stimulate cellular regeneration more effectively, promoting healthier skin cells and possibly slowing the skin aging process.

Protein Synthesis

Another critical aspect of skin health is protein synthesis. Proteins like collagen and elastin provide the skin with its elasticity and firmness. However, the body's ability to produce these proteins decreases with age, leading to a loss of skin elasticity and the formation of wrinkles.

In the study of short peptides, it was found that they control gene expression and the creation of proteins. These peptides can promote cell growth and division while inhibiting cell death (apoptosis), resulting in the recuperation of multiple organ functions. In addition to this, the majority of experiments demonstrated that peptides can boost the physiological capacity of cells, tissues, and the entire organism by up to 20-42%.

The administration of peptide-based medications in patients resulted in enhanced vitality and led to a decrease in mortality rates among the elderly and those of advanced age. 11]

Epitalon's influence on protein synthesis could potentially be beneficial for skin health. The peptide may enhance the body's ability to produce vital proteins such as collagen, thus helping to maintain the skin's elasticity and slow down the formation of wrinkles.

Cognitive Function

A significant area of Epitalon research relates to its effects on cognitive function. Neurogenesis, the formation of new neurons, is critical for maintaining cognitive abilities. As we age, this process naturally slows down, leading to cognitive decline. Epitalon has been shown to hold the capacity to manage the expression of genes associated with neuronal differentiation and the production of proteins in human stem cells through epigenetic modulation, suggesting it could potentially slow down age-related cognitive decline. 12]

Effects on Age-Related Diseases

One study found that Epitalon could potentially decrease the incidence of chromosome aberrations in senescence-accelerated mice. In a study of older, female mice (SAMP-1 mice), with signs of accelerated aging, more irregularities were found in their bone marrow cells than other types of mice. These irregularities, or "chromosome aberrations," were found to be decreased when the mice were treated with Epitalon from a young age.

Specifically, Epitalon reduced these irregularities by 20% in the SAMP-1 mice and by around 30.1% and 17.9% in two other types of mice, compared to mice of the same age who didn't receive the treatment. This suggests that Epitalon could help protect against the negative effects of aging. This finding suggests that the peptide may have protective effects against genetic damage associated with aging. 14]

Long-Term Studies on Epitalon

Research into the effects of Epitalon on gene expression is crucial to understanding its exact mechanisms. Studies have shown that long-term treatment with Epitalon can have specific effects on gene expression in mouse hearts, which corresponded with an increased lifespan in the mice. This suggests a possible mechanism through

58

which Epitalon could impact aging at a genetic level, possibly by influencing genes related to longevity and cellular repair. The aging process of the pineal gland, the organ responsible for melatonin production and circadian rhythm regulation, is another area where Epitalon's effects have been explored.

As we age, the activity of the pineal gland, which is responsible for producing the hormone melatonin, tends to decrease. This in turn leads to changes in our melatonin levels. Long-term observations suggested that certain substances, specifically Epithalamin derived from the pineal gland and the synthetic variant called Epitalon, can restore melatonin production in the pineal gland.

These substances also demonstrate a strong ability to regulate our body's neural, immune, and hormone systems, as well as its antioxidant activity. This supports the hypothesis that Epitalon might delay age-related changes in the body by preserving the functionality of crucial organs like the pineal gland. 15]

The Safety Profile of Epitalon

A significant part of the research into understanding the safety profile of Epitalon includes studying any possible adverse reactions reported in clinical trials and general observations.

The safety profile of a substance is a comprehensive account of its side effects, toxicity, interactions, and contraindications. This profile forms the basis for understanding the risk-benefit ratio of a potential therapeutic agent. For Epitalon, this profile is derived from numerous preclinical studies and limited clinical trials conducted so far. In preclinical studies, Epitalon exhibited a good safety profile. Animal models, including mice and Drosophila (fruit flies), have been used extensively to study the effects of Epitalon on lifespan extension and disease prevention.

In these studies, no significant toxic effects or adverse events were reported, even with long-term administration of the peptide.

Additionally, Epitalon's potential anti-cancer properties have been explored in studies involving transgenic HER-2/neu mice.

These studies found that long-term treatment with Epitalon suppressed the development of breast adenocarcinomas. This finding not only highlights Epitalon's potential anti-cancer properties but also supports its safety, given that it did not promote cancer development. Available reports suggest that Epitalon has also been well-tolerated in human subjects, with no significant adverse effects reported.

Actual adverse effects from Epitalon are remarkably rare. Some patients have reported experiencing certain issues during their regimen, however, it's unclear whether these symptoms were directly linked to Epitalon or simply coincidental, unrelated occurrences. Factors such as dosage, administration method, a patient's existing state of health, the duration of treatment, and individual patient characteristics can significantly influence tolerability and the occurrence of any side effects.

Limitations of Current Research on Epitalon

Epitalon has indeed generated significant interest due to its potential anti-aging properties. Although there are some limitations in the current research on Epitalon, it is crucial to remember that these limitations also represent opportunities for growth and new knowledge in this exciting field of study.

The size and scope of the studies conducted so far have laid the foundation for future research. Smaller studies and those involving animal models have been instrumental in establishing a basic understanding of Epitalon and its preliminary safety profiles.

These studies pave the way for large-scale, randomized controlled trials — the gold standard in clinical research — that can provide extended evidence on Epitalon's workings and efficacy.

While the diversity of participants in Epitalon studies has been somewhat limited, focusing primarily on specific populations such as the elderly or those with certain health conditions, this focus has allowed researchers to gain a deeper understanding of Epitalon's effects within these populations.

This knowledge can serve as a steppingstone for future studies involving a wider range of demographic groups, helping to broaden our understanding of Epitalon's potential applications. The current focus on Epitalon's potential benefits provides a strong basis for future studies to explore other aspects, such as factors involved in preventative healthcare and maintained vitality and wellness.

Future Research

In the field of anti-aging research, future research directions are aimed at exploring its potential benefits further. One avenue for future research involves the exploration of Epitalon's role in weight loss, muscle growth, and injury recovery. Given that these factors can significantly influence an individual's quality of life as they age, understanding how Epitalon impacts these areas could provide valuable insights into its potential as an anti-aging treatment.

Another promising area of research is the investigation of Epitalon's interaction with other therapeutic tetrapeptides. Studies using molecular dynamics simulation have compared complexes of lysine dendrimer and dendrigraft with therapeutic tetrapeptide Epitalon, providing a foundation for future research to build upon.

In addition, future research should also focus on investigating the sex differences in pharmacological interventions and their effects on lifespan and health span outcomes.

Considering gender-specific responses to anti-aging treatments like Epitalon could help optimize their use and effectiveness.

Preclinical research in cell lines and animal models has deepened our understanding of the molecular mechanism of various disease development.

Future directions should include further investigation into Epitalon's potential in drug interventions for conditions such as diabetic retinopathy. It's also worth noting that future research could explore Epitalon's reparative effect on the pineal gland ultrastructure, especially in gamma-irradiated rats.

This could open new avenues for understanding how Epitalon might mitigate the effects of radiation exposure, which is particularly relevant given the increasing utilization of radiation therapy in medicine. Lastly, the relationships between age-related changes in the biorhythms of the thymus endocrine function and pineal melatonin-producing function in healthy people could be another area for future research.

Understanding how Epitalon influences these functions could provide critical insights into how it might slow down or even reverse certain aspects of the aging process.

In conclusion, there exists substantial research on the broad effects of Epitalon, yet many potential therapeutic applications remain to be discovered. The areas identified for future research reflect the vast potential of this compound in anti-aging therapy and underscore the need for continued research into its workings, efficacy, and mechanisms of action.

CHAPTER 5

CARDIOVASCULAR BENEFITS

Cardiovascular health issues have become one of the leading causes of death in the United States, a reality that is deeply intertwined with our modern lifestyle. The Centers for Disease Control and Prevention (CDC) reports that heart disease is the leading cause of death for men, women, and individuals from most racial and ethnic groups in the US. Similarly, the World Health Organization (WHO) highlights that cardiovascular diseases (CVDs) are the leading cause of death globally, accounting for an estimated 17.9 million lives annually.

This alarming trend can be largely attributed to our modernized way of living, working, and eating. The nature of our contemporary lifestyle has drastically shifted from what it once was, and these changes have had profound effects on our health. Sedentary jobs, high-stress environments, insufficient sleep, and an over-reliance on processed foods have become commonplace in today's society.

Amid these lifestyle changes, another significant factor comes into play—aging. Cardiovascular health is significantly affected by the aging process. The impact of aging on the human body is profound and wide-reaching, affecting everything from our cognitive function to our cardiovascular health. As we age, our bodies undergo various changes, both externally and internally, that can significantly influence our overall health status.

The Relationship Between Aging
and Cardiovascular Health

Aging, a natural and inevitable process, involves numerous alterations at the cellular and molecular level, leading to functional changes in various organs and systems, including the cardiovascular system. As we age, the heart undergoes several anatomical and structural modifications. The most significant change is the weakening of the cardiac muscle or myocardium. This happens due to a combination of factors, including cellular senescence, apoptosis (programmed cell death), and reduced regenerative capacity of cardiac stem cells. Cellular senescence is a state where cells lose their ability to divide and function effectively. This loss of cell viability affects the heart's ability to contract since fewer cells are available to perform this function and pump blood efficiently.

Apoptosis often increases in aging hearts, leading to a decrease in the number of cardiac myocytes (muscle cells). The heart's regenerative capacity also diminishes with age, mainly due to a decline in the functionality and number of cardiac stem cells, further contributing to the weakening of the heart muscle. Aging also impacts the structure and function of blood vessels. The most noticeable effect is the stiffening of arteries, a condition known as arteriosclerosis. This occurs due to changes in the proteins (like collagen and elastin) that give the arterial wall its flexibility and due to the accumulation of advanced glycation end products (AGEs), which are harmful compounds formed when proteins or fats combine with sugar in the bloodstream.

As these AGEs accumulate on the arterial walls, they create cross-links between the molecules of the arterial wall, making it stiffer. Moreover, the smooth muscle cells in the arterial wall undergo changes to their characteristics, transitioning from a contractile state to a more synthetic state, which contributes to vessel stiffness.

64

The stiffening of arteries and arterioles due to aging adds significantly to the prevalence of high blood pressure in older adults. Stiffer arteries resist the flow of blood, causing the heart to work harder to pump blood, leading to an increase in blood pressure. Furthermore, atherosclerosis, characterized by the buildup of plaque inside the arteries, is influenced by aging.

The thickening of the inner layer of the artery wall provides an ideal environment for the accumulation of fats, cholesterol, calcium, and other substances that form plaque. Chronic inflammation, often associated with aging, damages the arterial lining, making it more susceptible to plaque buildup.

Common Cardiovascular Diseases Associated With Aging

While there are numerous age-related CVDs, some stand out due to their high incidence in older adults. These include coronary artery disease, heart failure, and arrhythmias. Each of these conditions has a unique way of manifesting and impacting overall health, but they all share a common thread—their link to the aging process.

Coronary Artery Disease (CAD):

Coronary artery disease (CAD), also known as coronary heart disease (CHD), is the most common type of heart disease in older adults. It occurs when the coronary arteries, which supply the heart muscle with oxygen-rich blood, become narrowed or blocked due to atherosclerosis. Aging contributes to the development of CAD through the progression of atherosclerosis. Over time, this plaque can harden or rupture, causing a complete blockage of the coronary artery, which then leads to a heart attack.

Symptoms of CAD may include chest pain (angina), shortness of breath, and fatigue. Oftentimes, however, there are no symptoms until a heart attack occurs. The impact of CAD on overall health is substantial as it can lead to heart failure, arrhythmias, and even death.

Heart Failure:

Heart failure, often a consequence of other CVDs like CAD, is another common condition in older adults. It occurs when the heart cannot pump enough blood to meet the body's needs. Age-related changes such as the weakening of the heart muscle, cellular senescence, and apoptosis contribute to the development of heart failure. With fewer viable cells available to contract, the heart's pumping capacity diminishes, leading to heart failure.

Symptoms of heart failure may include fatigue, shortness of breath, swelling in the legs, ankles, and feet (edema), and rapid or irregular heartbeat. Heart failure can severely impact a person's quality of life, making everyday activities challenging and leading to complications like kidney damage or failure.

Arrhythmias:

Arrhythmias refer to irregular heart rhythms, which become more common with age. They occur when the electrical signals that coordinate the heartbeats are not working correctly, causing the heart to beat too fast, too slow, or irregularly. Age-related changes in the heart structure, such as fibrosis (scarring of the cardiac tissue) and the decline in the number of pacemaker cells in the heart's natural pacemaker (the sinoatrial node), contribute to the development of arrhythmias.

Arrhythmias can cause a range of symptoms, from palpitations, chest pain, and shortness of breath to dizziness, fainting, and fatigue. Some arrhythmias can lead to debilitating stroke or sudden cardiac arrest if not treated promptly.

Effects Of Lifestyle On Cardiovascular Health In The Aging Population

The aging population faces an increased risk of cardiovascular diseases, largely due to physiological changes that come with age. However, lifestyle factors play a pivotal role in either worsening or reducing these risks. The four primary lifestyle factors affecting cardiovascular health are diet, exercise, smoking, and stress levels.

Diet

Diet plays a crucial role in maintaining cardiovascular health. Consuming foods high in saturated fats, trans fats, and cholesterol can lead to the buildup of plaques in our arteries. These plaques narrow the arteries and limit blood flow to the heart, thereby increasing the risk of heart disease. This process, known as atherosclerosis, can be accelerated by aging, putting older adults at a higher risk of developing CVDs.

Conversely, a balanced diet rich in fruits, vegetables, whole grains, lean proteins, and healthy fats can help maintain arterial flexibility, reduce inflammation, and manage body weight. All these factors contribute to better cardiovascular health and can slow down the progression of age-related cardiovascular issues.

Exercise

Regular physical activity is another key lifestyle factor influencing cardiovascular health. Exercise strengthens the heart muscle, maintains the elasticity of the arteries, and improves circulation. These benefits are particularly important for older adults, as they can help counteract some of the age-related changes in the cardiovascular system.

Moreover, exercise aids in managing body weight, reducing inflammation, and lowering blood pressure — all of which are essential for maintaining good cardiovascular health.

It also stimulates the production of new blood vessels and enhances the function of existing ones, providing additional support to the aging cardiovascular system.

Smoking

This is one of the most harmful lifestyle habits for cardiovascular health. The chemicals in tobacco smoke can damage the heart and blood vessels, leading to a range of cardiovascular diseases. In older adults, the effects of smoking can be even more detrimental. Smoking accelerates the stiffening of the arteries, a process that naturally occurs with aging. It also increases blood pressure, decreases good cholesterol (HDL), and promotes the formation of blood clots. All of these factors significantly increase the risk of cardiovascular diseases in the aging population.

Stress

Chronic stress is another lifestyle factor that can negatively affect cardiovascular health, especially in older adults. Long-term stress has been linked to high blood pressure, inflammation, and other heart disease risk factors. Over time, these effects can heighten the age-related changes in the cardiovascular system, leading to an increased risk of cardiovascular diseases. Additionally, people under chronic stress often resort to unhealthy behaviors such as smoking, overeating, or physical inactivity to cope, which can further compromise cardiovascular health.

How Epitalon Can Benefit Cardiovascular Health

Reducing Inflammation in Cardiovascular Tissues

One of the ways Epitalon is thought to benefit cardiovascular health is by reducing inflammation. Chronic inflammation is a key contributor to the development and progression of cardiovascular diseases. It can cause damage to the blood vessels, leading to hardening of the arteries (atherosclerosis), which can ultimately result in heart attacks or strokes. A study explored how Khavinson Peptides® (of which Epitalon is one) influenced certain molecules within our cells, specifically STAT1. STAT1, or Signal Transducer and Activator of Transcription 1, is a crucial protein found in our cells. Essentially, it can be likened to a type of command center that receives signals and then activates certain processes within the cell. When defending the body against viral infections, STAT1 is an important part of this process that helps mobilize the body's defenses.

These Khavinson Peptides® were found to be able to activate, or "turn on," STAT1. Moreover, it appears that these peptides can influence this process independently of certain receptors, noted by the regulation of a molecule called IFN-α in the study. IFN-α, or Interferon alpha, is a protein produced by our bodies particularly in response to viral infections. It's part of a larger group of proteins known as interferons, which play a key role in our immune system. IFN-α functions as a biological "alarm system" that signals other cells upon detecting a virus, activating their defense genes and helping to inhibit the spread of the infection. Given that STAT1 is commonly associated with our body's defense against viral infections, it raises the possibility that these peptides might imitate a similar effect. This could provide a new perspective on how we might leverage these peptides in medical applications. 16]

Epitalon may help reduce inflammation, thus lowering the associated risks with cardiovascular diseases. By decreasing the inflammatory response in cardiovascular tissues, Epitalon could potentially slow down the progression of atherosclerosis and other related cardiovascular diseases.

Improving Cholesterol Profiles

Another potential benefit of Epitalon is its role in improving cholesterol profiles. High levels of LDL cholesterol and low levels of HDL cholesterol—often known as "good cholesterol"—can increase the risk of heart disease. Epitalon may influence the body's lipid metabolism, helping to reduce levels of LDL cholesterol and increase levels of HDL cholesterol.

Research indicates that Epitalon significantly influences the regulation of the endocrine system in our bodies. The multitude of impacts that various hormones have on our bodily functions is critical, and a deficiency in these hormones can lead to various disease states. In the context of aging, there is a natural decline in hormone production due to the shortening of telomeres. Epitalon, however, has been shown to be able to enhance the endogenous levels of these diminishing hormones. 17] The endocrine system plays a pivotal role in maintaining the balance between High-Density Lipoprotein (HDL) and Low-Density Lipoprotein (LDL), which is crucial for optimum health.

The thyroid is a key gland in the endocrine system and produces different hormones that regulate our metabolism. If the thyroid is functioning well, it assists in controlling and breaking down LDL cholesterol, often referred to as the "bad cholesterol." Furthermore, hormones from a well-functioning endocrine system strengthen the production of HDL, or the "good cholesterol," which aids in the removal of LDL from the bloodstream. Therefore, a well-functioning endocrine system is integral to maintaining an optimal balance between HDL and LDL, contributing to overall cardiovascular health. This shift in the cholesterol profile may help prevent the buildup of plaques in the arteries, thus reducing the risk of heart disease.

Epitalon At The Cellular Level

Telomere Elongation

The shortening of our telomeres occurs in the heart just like all of the rest of our cells. By activating telomerase, Epitalon can extend the lifespan of the cells in the heart by counteracting the telomere shortening that would otherwise signal them to stop dividing and die. This ability to extend the lifespan of cells in the heart and blood vessels can potentially improve their resilience against age-related degeneration and stress. This, in turn, could lead to better overall cardiovascular function and health.

Antioxidant Properties

Epitalon's antioxidant properties stem from its ability to neutralize harmful free radicals and reduce oxidative stress within the cell. Free radicals are unstable molecules that can cause damage to cellular structures when their levels become too high, a condition known as oxidative stress. This stress is implicated in numerous health issues, including cardiovascular disease, neurodegenerative disorders, and aging. Free radicals are produced as byproducts of several biological processes, including metabolism. When their production exceeds the body's natural antioxidant defenses, it results in oxidative stress. This leads to damage to various cellular components, including DNA, proteins, and lipids, which can impair cell function and even lead to cell death.

Epitalon's role as an antioxidant involves hunting these free radicals and neutralizing them before they can cause harm. Specifically, it has been found to reduce lipid peroxidation, 2] a process where free radicals steal electrons from the lipids in cell membranes, resulting in cell damage. By reducing lipid peroxidation, Epitalon helps to protect the cell membrane and maintain its integrity and function.

71

In addition, Epitalon may also boost the body's natural antioxidant defenses by upregulating the production of antioxidant enzymes, such as superoxide dismutase and catalase. The use of the peptide Epitalon was found to greatly lower peroxide oxidation, often linked to cell damage (cutting it down to nearly one-fourth of the original amount). These enzymes play a vital role in neutralizing free radicals and preventing oxidative stress. 18]

In the context of cardiovascular health, the antioxidant properties of Epitalon could have significant benefits. Oxidative stress is a key factor in the development of cardiovascular diseases, contributing to inflammation and damage in the heart and blood vessels.

By reducing oxidative stress, Epitalon may help to prevent this damage. This could potentially reduce the risk of conditions like atherosclerosis and heart disease.

Evidence Of The Benefits
of Epitalon On Cardiovascular Health

As previously covered, research suggests that Epitalon could have beneficial impacts on cardiac health. In a review by Yiqiang Zhan and Sara Hägg, titled "Telomere length and cardiovascular disease risk" and published in the "Current Opinion in Cardiology" journal, the potential of telomere length as a biomarker for cardiovascular disease was explored. The authors discussed the biology of telomeres and their shortening capacity as an aging marker.

They also analyzed recent epidemiological evidence linking telomere length with cardiovascular risk factors and disease. The review highlighted the possible causal role of telomere shortening in coronary heart disease and summarized the potential biological mechanisms and pathways known. Current research suggested that short telomeres could be a risk factor for cardiovascular disease, possibly through insulin-mediated pathways.

However, further studies are needed to refine quantification methods, involve larger populations, and clarify the added role of telomere length in predicting future cardiovascular disease risks alongside existing biomarkers. 19]

Another study, titled "Optimization of cardiovascular function by peptide bio-regulators," by V. A. Cherkashin, G. F. Semin, and A. A. Veretenko, explored the effects of peptide bioregulators. The effect of thymalin, Epitalon (referred to as Epithalamine), and cortexine on systemic hemodynamics in patients with cardiovascular and cerebrovascular conditions was examined.

The study assessed the functional stability and the quality of cardiovascular system regulation based on the principles of automatic regulation theory. The results showed that these peptide bioregulators were highly effective in improving the hemodynamic parameters of the patients studied. 20]

Another study titled, "Geroprotective effect of Epithalamine (pineal gland peptide preparation) in elderly subjects with accelerated aging" by O.V. Korkushko, V. Kh. Khavinson, V.B. Shatilo, and I.A. Antonyuk-Shcheglova, published in the Bulletin of Experimental Biology and Medicine, was conducted.

This was a 12-year randomized clinical trial on elderly patients with coronary disease, and accelerated cardiovascular aging. The study found that long-term treatment with Epithalamine decreased the functional age and degree of cardiovascular aging and increased exercise tolerance.

Notably, after 12 years, the mortality rate was 28% lower in the group treated with Epithalamine compared to the control group, despite both groups receiving the same basic therapy. Moreover, cardiovascular mortality was two times lower in the Epithalamine-treated group.

And the incidence of cardiovascular failure and respiratory diseases was also two times lower. The study concluded that long-term treatment with Epithalamine had a geroprotective effect, improving the long-term life prognosis in elderly subjects with accelerated aging. 21]

CHAPTER 6

NEUROPROTECTIVE PROPERTIES

A WARNING FOR MY READERS!

This chapter is very technical and has a lot of brain science in it. If you are not inclined to like a lot of scientific data and names of biological processes, peptides, pathways, etc., the details of this chapter might be a little too much for you.

I'll summarize: Epitalon is very good for helping maintain and replace neurons in the brain, and there is promising research regarding this! It has the potential to help prevent and reverse age-related brain diseases without side effects! MUCH more research needs to be done, but the outlook for the future regarding Epitalon and the brain is bright!

In the ordinary course of life, our cognitive abilities are fundamental to the overall functionality and quality of living. These abilities govern everything, from the way we manage information to how we make choices and how we retain instructions or engage with our environment. They also govern the way we solve problems (or our capability to do so) and how well we manage stress, or to what degree the environment appears overwhelming to us.

Yet, as we age, there appears to be a "natural" reduction in our cognitive abilities, impacting our capacity to perform even the most basic of activities. And, if we happen to carry the genes leading to neurodegeneration, it may seem as if it is entirely up to fate whether we lose our ability to even remember who we are as we grow older, continually meeting a new person in our own house throughout the day. Referred to as cognitive decline, this phenomenon has become a noteworthy health issue, especially in nations like the United States where there is a growing elderly population.

According to data from the Centers for Disease Control and Prevention (CDC) in 2011, more than 16 million people aged 18 years or older in the U.S. were living with some form of cognitive impairment. 22] The exact causes of cognitive decline are multifactorial and typically include both physical and mental health factors. Physical health issues such as stroke, heart disease, diabetes, and hypertension can all increase the risk of cognitive impairment. Mental health conditions, like depression and anxiety, have also been known to contribute to cognitive decline. In fact, some studies have even suggested that chronic stress can lead to changes in the brain that affect memory and cognitive function. 23]

Yet conventional medicine offers no known cure for common neurodegenerative conditions such as Alzheimer's Disease or Parkinson's Disease, aside from symptom management. Therefore, aging still remains one of the most significant risk factors for cognitive decline, with the inevitable probability of suffering neurodegeneration increasing as we age. However, what if there were natural and powerful ways to not only counteract neurodegeneration but altogether prevent it?

Common Age-Related Diseases and Their Symptoms

Some changes that come with age include an actual decrease in brain volume. This refers to a reduction in blood flow to the brain and the accumulation of harmful proteins that may lead to neurodegenerative diseases. These can significantly affect an individual's quality of life and may require long-term care or management without any known cure.

There are estimates that we lose about 9,000 neurons per day. Estimates range widely depending on who reports this, but we normally only replace somewhere between 2,000-4,500 neurons per day!

Alzheimer's Disease

Alzheimer's Disease is a notoriously insidious condition that gradually erodes a person's cognitive ability over time. This often begins with subtle changes, such as difficulty remembering recent conversations or events. This is often initially mistaken for normal forgetfulness connected to just growing older. However, as the disease progresses, memory loss becomes more pronounced, extending to essential personal information such as the names of close family members, significant life events, where the person lives, or even their own name.

In addition to memory loss, individuals with Alzheimer's Disease may also experience difficulties with basic problem-solving skills and complex tasks. For example, they may struggle with managing their finances, planning their own meals, or following a basic recipe when cooking. In extremely progressed Alzheimer's, the person loses basic skills such as tying their shoelaces. Language skills can also be affected, with a person having trouble finding the right words in a conversation or understanding written text. Behavioral changes are quite common too, as the individual's personality is said to "erode." This can manifest as mood swings, social withdrawal, or personal characteristics that are unusual compared to their normal behavior, like apathy or aggression.

An example could be an elderly woman, let's say a 75-year-old retired teacher. Her family might notice that she was beginning to frequently misplace her keys and forget scheduled appointments. She might repeat simple questions multiple times, seemingly unaware that they had already been answered.

As time progresses, she may begin to struggle with cooking meals she normally knows how to prepare—a task she has previously enjoyed and excelled at. She could suddenly forget to turn off the stove or become confused about the steps involved in her favorite recipes. These changes would, of course, be distressing for both the person and the family.

The underlying mechanism for these symptoms in Alzheimer's disease revolves around the fact that abnormal protein aggregates accumulate in the brain. In the brains of Alzheimer's patients, two types of protein aggregates are observed to accumulate:

Amyloid-beta plaques: These are sticky clumps of protein fragments that accumulate outside of the brain's nerve cells. Amyloid-beta is a protein fragment that is produced by normal brain activity. In healthy individuals, these fragments are broken down and eliminated. However, in Alzheimer's patients, the fragments accumulate and form hard, insoluble plaques.

Tau tangles: These are twisted threads of protein found within nerve cells. The tau protein helps maintain the structure of a neuron, including its transport system. In Alzheimer's disease, the tau protein changes shape and organizes itself into structures called neurofibrillary tangles. These tangles disrupt the transport system and lead to the death of the nerve cells. These proteins disrupt the normal functioning of neurons, leading to their death. Over time, this neuronal death spreads throughout the brain, causing the observed decline in cognitive function. 24]

Amyloid-beta plaques are sticky buildups that accumulate outside of the neurons.

They are thought to block neuronal communication, triggering an inflammatory response that leads to damage and death of the neurons. Tau tangles, on the other hand, are a form of twisted fibers that form inside the neurons. In healthy neurons, tau proteins help maintain the structure of microtubules, which are essential for nutrient transport within the neuron.

However, in Alzheimer's disease, these tau proteins become abnormal and clump together, forming tangles that disrupt this transport system and lead to neuronal death. These pathological changes in the brain usually begin years—even decades—before the first symptoms of Alzheimer's appear. This certainly highlights the insidious nature of this condition and the need for effective preventative measures.

Parkinson's Disease

Parkinson's disease is another neurodegenerative disorder that primarily affects bodily movement, resulting in symptoms that can significantly impact daily life. The characteristic symptoms of Parkinson's disease are often grouped into motor and non-motor symptoms.

Motor symptoms:

Motor symptoms include manifestations such as resting tremors, muscle stiffness or rigidity, bradykinesia (slowness of movement), and postural instability. Resting tremors usually start in one hand and may progress to affect the entire arm. This could manifest as difficulty performing simple tasks, like holding a cup of coffee, due to uncontrollable shaking or tremors.

Muscle stiffness can make it very difficult for individuals to perform routine tasks, such as buttoning a shirt or tying shoelaces, as well as causing great discomfort and pain. Bradykinesia, on the other hand, makes everyday activities such as walking or eating slow and difficult. Postural instability, which refers to balance problems and frequent falls, truly compromises an individual's safety.

Non-motor symptoms:

Non-motor symptoms include cognitive changes, mood disorders (like depression and anxiety), disturbances in sleep patterns, constipation, and even a loss of smell. These symptoms can be quite debilitating and often precede the motor symptoms by years.

Consider the case of a 68-year-old retired engineer. He might first begin noticing a persistent tremor in his right hand, which is more pronounced when he is resting and lessens with movement. Over time, his family starts to observe that his movements have slowed and that he often seems rigid in his movements. His handwriting also becomes smaller and "cramped," a condition known as micrographia. He may later experience difficulty with maintaining his balance, which leads to occasional falls. He also reports feeling depressed and anxious and starts having trouble sleeping at night.

The root cause of all these symptoms in Parkinson's disease is the loss of dopamine-producing cells in a region of the brain called the substantia nigra. Aside from being known as the "feel good" hormone, dopamine is also a neurotransmitter used to regulate movement and coordination. When the substantia nigra cells begin to die off, dopamine levels decrease.

This disrupts the balance of activity between two other brain regions — the basal ganglia and the thalamus. This imbalance leads to the motor symptoms characteristic of Parkinson's disease.

Moreover, the non-motor symptoms suggest that Parkinson's disease is not just a movement disorder but also involves a broader range of brain changes that affect multiple brain regions and neurotransmitter systems.

Vascular Dementia

Vascular dementia, as the name suggests, is a type of cognitive decline that results from reduced blood flow to the brain.

This often occurs following a stroke or a series of mini-strokes (transient ischemic attacks), leading to a stepwise decline in cognitive function. This is different from the gradual decline typically seen in Alzheimer's disease. Symptoms of vascular dementia can vary greatly depending on the area and extent of the brain affected by the reduced blood flow. Common symptoms include memory loss, confusion, and difficulties with concentration and decision-making.

Other possible symptoms include physical weakness or paralysis, difficulty walking, problems with balance and coordination, mood changes, and urinary incontinence.

To illustrate, let's consider the case of a 72-year-old woman with a history of high blood pressure and heart disease. One day, she suffers a minor stroke. Afterwards, her family notices significant changes in her behavior. She begins to struggle with tasks that require planning and decision-making, such as managing her finances or organizing a grocery list. She frequently begins to forget about recent events and has difficulty following conversations.

Her mood also fluctuates, with periods of depression and apathy. These changes are a clear departure from her previous behavior, causing her loved ones great concern.

The underlying mechanism of vascular dementia involves damage to the brain's blood vessels due to conditions like high blood pressure, diabetes, and high cholesterol, or events like a stroke. When these blood vessels are damaged, they may narrow, rupture, or even leak, reducing the flow of blood to the brain. This can lead to areas of the brain not receiving enough oxygen and nutrients, causing brain cells in these areas to become damaged or die. This cell death can lead to the cognitive and physical symptoms observed in vascular dementia.

Standard Treatment

The cognitive diseases mentioned earlier are complex conditions that typically require a multifaceted treatment approach. The goal of most of these treatments is not only to manage symptoms and slow disease progression but also to try and improve the quality of life for those affected.

Alzheimer's Disease

The standard medical treatment for Alzheimer's disease includes certain drugs such as cholinesterase inhibitors (ChEIs) like Donepezil, Rivastigmine, and Galantamine.

These medications work by preventing the breakdown of acetylcholine, a neurotransmitter that is important for memory and learning, thereby increasing its availability in the brain. Another medication used is Memantine, a partial N-methyl-D-aspartate (NMDA) antagonist that works by regulating the activity of glutamate, another neurotransmitter involved in brain cell communication.

While the most common medications for Alzheimer's can manage some of the symptoms, they do not offer a cure for the condition. They may biochemically help to temporarily improve memory, attention, reason, language, or the ability to perform simple tasks, but also come with some unwanted side effects.

Donepezil:

The most common side effects of Donepezil include nausea, vomiting, a loss of appetite, and increased frequency of bowel movements. Serious side effects can include a slow heart rate, fainting, and even seizures. Additionally, some people might experience trouble sleeping, muscle cramps or weakness, and shakiness (tremors).

Rivastigmine:

This drug also has side effects that include nausea, vomiting, loss of appetite, and dizziness. More serious side effects can include stomach bleeding, seizures, and worsening lung problems if a person suffers from asthma or other disorders connected to the lungs.

Galantamine:

Side effects are similar to the ones above, with the addition of weight loss, headache, and stomach pain. More serious side effects can include a very slow heart rate, bloody or tarry stools, coughing up blood, and decreased urination.

Memantine:

Common side effects include dizziness, confusion, and headache. Serious side effects can include shortness of breath, hallucinations, severe dizziness, and a fast heartbeat.

Parkinson's Disease

The primary conventional treatment for Parkinson's disease is Levodopa. This is a class of drugs called central nervous system agents. The brain converts Levodopa into dopamine to replenish the depleted supply. Other medications include dopamine agonists, which mimic the effects of dopamine, and MAO-B inhibitors, which help prevent the breakdown of dopamine in the brain. However, these drugs also come with an array of side effects:

Levodopa:

Although this drug is often effective at reducing or eliminating the tremors and other motor symptoms of Parkinson's disease during the early stages of the disease, after prolonged use, it can cause side effects. These include nausea, vomiting, low blood pressure, and restlessness. More serious side effects include confusion, hallucinations, and unusual urges like increased sexual urges, gambling, etc. Another significant issue with Levodopa is "wearing-off," where the medication's effects wear off between doses, causing symptoms to return before it's time for the next dose.

Dopamine Agonists:

The common side effects of these drugs often include nausea, orthostatic hypotension (a sudden drop in blood pressure when standing up), sleepiness, and compulsive behaviors like gambling, increased sexual behavior, overeating, or shopping. They may also cause hallucinations and psychosis, particularly in older people.

MAO-B Inhibitors:

Side effects can include insomnia, dizziness, nausea, and dry mouth. More serious side effects can include high blood pressure (when taken with certain foods or medications) and hallucinations.

Vascular Dementia

On the other hand, treatments for vascular dementia often focus on trying to manage the underlying conditions that affect the blood vessels, such as high blood pressure, diabetes, and heart disease. This may involve medications like antihypertensives, statins, anticoagulants, or antiplatelets. However, as with these other conditions mentioned above, this treatment does not cure the condition itself and many of these drugs come with their own side effects:

Antihypertensives:

These medications are used to control high blood pressure. Side effects can include dizziness, feeling tired, upset stomach, and dry cough. More serious side effects can include kidney problems, a high potassium level, and allergic reactions.

Statins:

These drugs are used to lower cholesterol levels in the blood. Common side effects include headache, difficulty sleeping, skin flushing, muscle aches, nausea, bloating, or gas. Serious side effects can include liver damage, increased blood sugar or type 2 diabetes, and neurological side effects like memory loss or confusion.

Anticoagulants:

They prevent blood clots from forming and reduce the risk of stroke. Side effects can include excessive bleeding, hemorrhagic stroke, and gastrointestinal bleeding. Other side effects can include dizziness, weakness, bruising, and hair loss.

Antiplatelets:

These drugs reduce the risk of heart attacks and strokes by preventing blood clots. Side effects can include bleeding more easily, rash, or diarrhea. More serious side effects can include stomach ulcers, abdominal pain, and bloody or tarry stools.

However effective these conventional treatments may be at immediately managing or suppressing the symptoms manifesting themselves, it is clear that we are in need of another approach. A more effective strategy would be one where we could prevent—without causing any other harm—the development of neurological conditions, or repair and heal them in a holistic manner.

Epitalon and Cognitive Diseases

Ongoing research in the field of neuroscience and molecular biology focuses on developing safer and more effective treatments, offering hope for better management of cognitive diseases in the future. One intriguing area of study revolves around the synthetic peptide Epitalon.

As previously mentioned, Epitalon has piqued scientific interest due to its potential anti-aging properties.

In addition to its anti-aging properties, however, Epitalon has demonstrated potential in regulating various organ functions, including those in the pineal gland, retina, and brain. This broad impact on bodily functions could culminate in numerous health benefits, such as improved sleep quality and increased energy levels.

In the context of cognitive diseases, Epitalon's ability to regulate brain function and slow cellular aging could be particularly beneficial. It is important to note, however, that while early results are promising, more extensive clinical trials are needed to fully understand Epitalon's effects and safety profile.

The exploration of Epitalon and similar compounds represents a significant step forward in our understanding and treatment of cognitive diseases. As we delve deeper into these innovative therapies, we inch closer to a future where cognitive diseases can be effectively managed, improving the quality of life for millions worldwide. To fully understand the importance and potential of Epitalon, it's crucial to delve into three key aspects of brain function and aging: neurogenesis, neural protection, and cellular aging.

Neurogenesis is the process of forming new neurons or nerve cells. As we age, neurogenesis decreases, which could lead to poorer performance on learning and memory tasks and a reduced ability to distinguish smells. Epitalon's potential to stimulate neurogenesis could be a significant beneficial change for cognitive health in older adults.

Neural protection refers to the various mechanisms that protect neurons from damage and death. New neurons, especially during neurogenesis and neuronal maturation, are hypersensitive to telomere and genome damage.

Here, again, Epitalon could play a vital role by enhancing the body's natural neural protection mechanisms.

Cellular aging is an integral part of the aging process. As mentioned earlier, each cell division results in the shortening of telomeres. When telomeres become too short, the cell can no longer divide and becomes senescent or dies. This process contributes to the overall aging process and the decline in brain function. By stimulating the production of telomerase, Epitalon could potentially slow down cellular aging.

Epitalon vs. Other Treatments

With its distinctive longevity-enhancing effects and neuroprotective properties, Epitalon is a peptide that presents us with an intriguing alternative to traditional treatments for cognitive disorders.

Neuroprotective Properties

Another aspect where Epitalon shows its potential is through its neuroprotective properties. This property is particularly important as oxidative stress and DNA damage increase with natural aging, which in turn contributes to cognitive decline and various neurodegenerative diseases. Most conventional drugs used as treatment for cognitive decline, such as cholinesterase inhibitors for Alzheimer's Disease, function by seeking to increase the levels of certain chemicals in the brain to assist memory and cognitive function. However, none of these drugs offer any neuroprotection, nor do they address the underlying neuronal damage taking place.

Epitalon's Effect on Memory Enhancement

Memory is an essential part of our cognitive function that is crucial in forming our daily experiences and quality of life. Research conducted by Khavinson et al. in 2020 provides compelling evidence of the potential role of Epitalon (AEDG peptide) in enhancing memory function. The study reveals that Epitalon influences neurogenic differentiation gene expression and protein synthesis in human gingival mesenchymal stem cells, suggesting an epigenetic mechanism for this process. Epitalon was found to regulate the function of the pineal gland, retina, and brain—key areas involved in cognitive processes, including memory. Its observed ability to induce neuronal cell differentiation in retinal and human periodontal ligament stem cells further emphasizes its potential role in neurogenesis, a process crucial for memory formation and consolidation.

In terms of specific impacts on memory-related proteins, Epitalon was shown to increase the synthesis of important neurogenic differentiation markers such as Nestin, GAP43, β Tubulin III, and Doublecortin in human gingival mesenchymal stem cells. These proteins play vital roles in the growth and development of neurons, which are essential for memory function.

Furthermore, the study found that Epitalon can bind with certain histones (H1/6 and H1/3) in specific sites that interact with DNA. This interaction could potentially lead to an increase in neuronal differentiation gene transcription of Nestin, GAP43, β Tubulin III, and Doublecortin. This suggests that Epitalon may epigenetically influence the expression of genes involved in neuronal differentiation and, hence, memory enhancement. 4]

Similarly, a study by Ilina et al. explored the neuroepigenetic mechanisms of ultrashort peptides, like Epitalon, in Alzheimer's disease. Epigenetic regulation of gene expression is essential for preserving higher cognitive functions like learning and memory. The current understanding of epigenetics in Alzheimer's disease revolves around DNA methylation, chromatin remodeling, histone modifications, and regulation of non-coding RNAs.

The pathological links of Alzheimer's disease include misfolding and aggregation of tau protein and amyloid peptides, mitochondrial dysfunction, oxidative stress, impaired energy metabolism, destruction of the blood-brain barrier, and neuroinflammation.

These factors contribute to impaired synaptic plasticity and memory loss. Ultrashort peptides have promising neuroprotective properties with a wide range of activity and no reported side effects. It was theorized that peptide regulation of gene expression could be mediated by the interaction of short peptides with histone proteins, cis- and trans-regulatory DNA elements, and effector molecules (DNA/RNA-binding proteins and non-coding RNA). 25]

The Activation of Chromatin

The process of creating and stabilizing memories is closely linked to the intricate management of gene activity, steered by the organization of chromatin. Chromatin is the material within our cells that packages DNA into a more compact, denser form. It plays a crucial role in gene expression, DNA replication, and repair.

During the phases of memory stabilization, specific genes are selectively switched on, while others are dialed down. Such selective activation or repression of genes is made possible by adjustments in the chromatin architecture, facilitated by chemical processes including the methylation and acetylation of histones. Histones are a family of proteins that associate with DNA in the cell to form nucleosomes, which are the fundamental subunits of chromatin.

These proteins not only function as "spools" around which DNA winds but also play a pivotal role in regulating the accessibility of genetic information by undergoing various chemical modifications, thus influencing gene expression. Chromatin alterations are also central to the brain's remarkable capacity for neuroplasticity — its ability to reshape itself in reaction to various stimuli.

DR. JON HARMON, DC, BCN, ICP

This restructuring of chromatin can prompt a cascade of gene expression events, pivotal for the development, strengthening, or even removal of synaptic connections. This mechanism, however, when disrupted, can be a contributory factor in an assortment of brain-related conditions. Deviations in chromatin dynamics have been noted in diverse neurodevelopmental and neurodegenerative diseases, including but not limited to Rett syndrome, Fragile X syndrome, and Alzheimer's disease. Mutations that affect enzymes that modify chromatin or the structural aspects of chromatin itself can induce extensive gene expression changes, thereby playing a significant role in the manifestation of these disorders. 26]

Epitalon has been demonstrated to activate chromatin by modifying heterochromatin and heterochromatinized chromosome regions in cells of the elderly. This modification could potentially lead to the upregulation of genes that are beneficial for health and longevity, aside from how it might benefit and promote cognitive function through the activation of chromatin.

Neurogenesis

Research indicates that Epitalon is instrumental in encouraging human periodontal ligament stem cells to differentiate into neurons. This is a crucial phase within the larger scope of neurogenesis, or the creation of new neurons. These nascent neurons are poised to merge with preexisting neural circuits, potentially boosting the brain's adaptability and cognitive performance. 27]

The term "nascent neurons" refers to newly formed neuronal cells that have recently undergone cell division but have not yet fully matured to acquire distinct neurological functions. These fledgling cells hold the potential to integrate into existing neural circuits and contribute to cognitive processes once they have matured.

This distinguishing attribute of Epitalon might offer protection against neurodegenerative conditions marked by the gradual deterioration of neuronal health. Through fostering neuron development and maturity, the AEDG peptide could potentially offset the decline in neurons, thereby decelerating cognitive regression. Epitalon's influence on the genes governing circadian rhythms may also be of interest, as disruptions in this balance have been linked to cognitive deficits and neurodegenerative brain conditions.

Furthermore, Epitalon's positive effect on our visual health, by prompting retinal stem cells to engage in neuronal differentiation, could support cognitive health as the retina is an integral part of the central nervous system in relation to the brain. Due to the connection between retinal degradation and an increased risk for cognitive decay, the promotion of retinal integrity could mean improved brain function and enhanced cognitive soundness.

The implications of this research on memory improvement are significant. By potentially regulating the gene expressions related to neuronal function and protection, Epitalon could help maintain synaptic plasticity, which is crucial for memory formation and retention! This implies that Epitalon and other ultrashort peptides could be key components in a multifunctional treatment of Alzheimer's disease and many other memory-related disorders.

Epitalon and Neural Health

The future of Epitalon research in the field of neural health is evolving in several exciting directions. One area of interest is understanding the epigenetic mechanism through which Epitalon stimulates gene expression and protein synthesis during neurogenesis. Research advancements in aging and associated diseases that particularly focus on brain aging and brain health are another promising direction. Future studies could explore how

Epitalon influences brain aging and whether it can support a healthier brain as we age.

Another fascinating avenue for future research lies in exploring the role of Epitalon in bioregulation. This could help in the development of new therapeutic applications and the possibility of using Epitalon to estimate our biological age, based on the spectral analysis of the bioelectrical activity of the human brain, could open doors for new research. There is so much more to be discovered and enhanced, from understanding various epigenetic mechanisms and their role in neurogenesis to exploring the impacts of Epitalon on neuronal disease prevention. The potential of Epitalon in bioregulation, its role in correcting stress-induced dysfunctions, and its ability to normalize epigenetic changes are all promising areas for future research.

With new clinical trials and milestone developments, future research into Epitalon's mechanisms could unlock significant breakthroughs in neural health.

CHAPTER 7

IMMUNE SYSTEM FORTIFICATION

Aging is a natural and inevitable process that all living organisms go through. However, aging alters our physiology far more than the visible changes we often identify with aging. Far beyond wrinkles, fine lines, and graying hair, the aging process greatly impacts our bodily systems and cellular functions. Among the various systems severely affected by aging is our immune system. This complex network of cells, tissues, and organs, which is so vital to maintaining a healthy state and protecting us from pathogens, undergoes a gradual decline in functionality as we age.

This process, known as immunosenescence, significantly affects our ability to fight off infections and diseases. The implications of this decline are wide-ranging, influencing not only our susceptibility to illnesses but also our overall health and well-being. The complex process of immunosenescence essentially leads to a decline in the body's ability to respond to infections and to develop robust immunity.

As we age, key immune cells, such as T-lymphocytes, become less effective. Furthermore, the thymus, which is responsible for the maturation of these cells, shrinks and produces fewer T cells.

Additionally, aging immune systems tend to show a reduced diversity of B cells, which play a critical role in producing antibodies. Consequently, older adults are at a higher risk of contracting infections, experience diminished healing, and often suffer from more severe and prolonged illnesses.

Immunosenescence also contributes to the increased incidence of autoimmune disorders and cancer in the elderly, as the immune surveillance that typically helps to eliminate cancer cells and regulate immune responses becomes compromised. Given the breadth of the impact that aging has on our immune system, it becomes apparent that any measures we could employ, which enhance the body's inherent capabilities to bolster our immune function, would yield considerable benefits! By reinforcing our body's natural defenses, we improve our chances of warding off disease and maintaining better overall health as we age.

What Happens to Our Immune System as We Age?

The phenomenon of immunosenescence includes thymic involution, altered T and B cell responses, and an altered ratio of naïve immune cells (essential components of the immune system that enable the body to fight off new, unrecognized infections and diseases) to memory cells. Thymic involution is a process in which the thymus, a primary lymphoid organ situated in the upper anterior portion of the chest cavity, gradually shrinks and loses functionality over the course of an individual's lifespan. Initially peaking in size and activity during childhood, the thymus is critical for the development and maturation of T-lymphocytes or T-cells, which are vital components of the adaptive immune system.

As one ages, the functional tissue of the thymus is progressively replaced by fatty and fibrous tissue, resulting in a decreased output of new T-cells. This can in turn contribute to a weakened immune response in older adults. This process is influenced by various factors, such as hormonal changes and stressors. As we age, our immune function becomes impaired, which can lead to more severe infections. This is due to various factors, like the remodeling of lymphoid organs which then leads to changes in the immune system's structure and overall degree of functioning.

The immune system's gradual deterioration due to natural age advancement is a significant factor in immunosenescence.

One of the primary changes that occur with aging is a decrease in the production of white blood cells, also known as leukocytes. White blood cells are the body's primary defense against infection and essential agents in the immune response. They are responsible for identifying, attacking, and eliminating pathogens. However, as we age, the bone marrow's ability to produce these cells diminishes, leading to a decreased number of circulating white blood cells. This reduction can leave the body more susceptible to infections and illnesses.

Inflamm-Aging

Another aspect of aging and the immune system is the concept of "inflamm-aging." This refers to a low-grade, chronic inflammatory state that is often present in older adults. Our contemporary lifestyle, marked by prolonged sitting, processed foods, a toxic environment, and high levels of chronic stress, induces inflamm-aging in ways that extend far beyond the expected, natural wear and tear of merely growing older.

Our environment and way of living significantly contribute to the degree of systemic inflammation observed in today's aging population. Systemic inflammation refers to a widespread immune response to perceived threats within the body, such as infection or injury. Rather than being confined to a specific area, this inflammatory response occurs throughout the entire body and impacts multiple organs and tissues. It is characterized by the release of inflammatory cytokines into the bloodstream, which contribute to chronic health conditions if persistently elevated or uncontrolled.

Cytokines are a broad category of small proteins that are important in cell signaling. They are released by cells and affect the behavior of other cells and often play a pivotal role in the immune system in mediating and regulating immunity, inflammation, and hematopoiesis. Hematopoiesis is the process through which the body manufactures blood cells, including the development of all cellular blood components in the bone marrow. This process involves the differentiation of multipotent hematopoietic stem cells into various types of blood cells, including red blood cells, white blood cells, and platelets, which are essential for carrying oxygen, fighting infections, and blood clotting, respectively.

Inflammatory cytokines are essentially signaling proteins that are released by cells, particularly immune cells, during systemic inflammation. They serve as messengers to mediate and regulate immunity, inflammation, and hematopoiesis. In the context of systemic inflammation, they can however activate a cascade of detrimental biological responses.

When released excessively, these cytokines lead to a state known as a "cytokine storm" which can damage tissues and organs, disrupt normal cellular function, and in severe cases, be life-threatening.

Chronic, low-grade exposure to these cytokines is known to contribute to the development of diseases like atherosclerosis, diabetes, and fibrosis, by sustaining an environment of persistent inflammation.

These factors, coupled with the frequent exposure to toxins and chemicals found in modern materials and foods, provoke a persistent inflammatory response. Additionally, social isolation and sleep deprivation, two other hallmarks of our modern life, are increasingly recognized as contributors to inflamm-aging. Unlike the occasional stressors faced by our ancestors, the chronic stress of the 21st-century lifestyle keeps the body in a state of being constantly "on edge," which further tips the balance towards a pro-inflammatory state.

This constant stress level also keeps a person in "sympathetic freeze" which is a state in which the sympathetic nervous system (fight or flight) is forever in a state of activation. This inhibits the parasympathetic nervous system! These are the two parts of the autonomic nervous system and, unfortunately, it is the parasympathetic portion of the autonomic nervous system that is responsible for the "anti-inflammatory" function! When a person is in a state of "sympathetic freeze," the anti-inflammatory pathways are inhibited which contributes to this "inflamm-aging!"

Aging and the Ability to Fight Diseases

As mentioned, as our immune system's efficiency tends to decrease with age this not only results in a heightened risk of infection, delayed wound recovery, but also in an increased occurrence of autoimmune diseases. One of the most common health issues among older adults is an increased frequency of respiratory infections.

The lungs, like every other part of the body, are not exempt from the effects of aging. With age, the elasticity of the lungs decreases and lung muscle strength declines, weakening the immune defense of the lungs. These changes can make it harder to breathe and increase one's vulnerability to respiratory infections like pneumonia and influenza. Aside from this, a weakened immune system not only does not respond as effectively to respiratory infections but may also lead to prolonged and difficult recovery times.

Another area where the effects of aging are evident in our immune system is wound healing. The healing process itself (the Healing Cascade) involves several phases, such as inflammation, tissue formation, and tissue remodeling, all of which require a robust immune response. As we age, however, this process slows down due to factors such as reduced skin elasticity, decreased cell division, and a less efficient immune response.

Therefore, wounds in older adults tend to take longer to heal, have a higher risk of infection, and often result in more significant scarring. The skin's immune response also changes with age. The skin itself is the body's first line of defense against pathogens and it contains specialized immune cells that help combat infections.

However, with age, the skin becomes thinner and tends to be drier. The number of immune cells in the skin also decreases. These changes can compromise the skin's barrier function, making it easier for pathogens to enter the body and cause infections.

Epitalon and Our Immune System

An aging immune system, with its reduced ability to fight off infections and slow wound healing, has been a focus of numerous studies and research. Several treatments have been proposed to mitigate these effects and improve immune function as we grow older.

One such approach is the use of mammalian Target of Rapamycin (mTOR) inhibitors, as suggested by Mannick et al., 2018. The mTOR is a protein that regulates cell growth, proliferation, motility, and survival. By inhibiting mTOR, it is possible to enhance the immune response and increase resistance to infectious diseases in older adults. 29]

Another strategy has been to rejuvenate aged immune tissues or cells through techniques like stem cell therapy, where new and healthy cells are introduced to replace or repair damaged or aged cells. Similarly, immunosuppressive strategies have also been used to try and suppress dysregulated immune responses and reduce the risk of autoimmune disorders where the immune system mistakenly attacks the body's own cells.

However, Epitalon has also gained a lot of recent attention for its potential benefits on the immune system. This peptide has been shown to have a wide range of positive effects on immune function such as lymphocyte activation, antibody production, and cytokine regulation.

Lymphocyte Activation and Antibody Production

Lymphocytes are a central component of the adaptive immune system, which is the part of the immune system that learns and adapts to new threats. They are categorized into two main types: B lymphocytes (B cells) and T lymphocytes (T cells). Both play critical roles in the immune response, but they do so in different ways:

B cells primarily function by producing antibodies. These are proteins that can specifically recognize and bind to antigens which are substances that the immune system recognizes as foreign. This binding neutralizes the antigen and tags it for destruction by other immune cells.

T cells, on the other hand, come in several varieties, each with its own function. Helper T cells stimulate B cells to produce antibodies and help activate killer T cells. Killer T cells, also known as cytotoxic T cells, can directly kill off cells that have been infected with viruses or have otherwise become cancerous.

The process of lymphocyte activation is a vital one in the body's immune response. It involves a complex series of cellular events that begin when a lymphocyte encounters an antigen. When an antigen enters the body, it encounters a sophisticated and multi-layered immune system. Initially, antigen-presenting cells (or APCs), such as dendritic cells, recognize the antigen as a foreign entity. These APCs then engulf the antigen and "process" it, presenting fragments of the antigen on their surface coupled with major histocompatibility complex, or MHC, molecules.

This presentation is a critical part of the immune response as it enables T cells, specifically helper T cells, to recognize and bind to the antigen-MHC complex through their T cell receptors (TCR).

This interaction is highly specific and triggers the T cells to activate and proliferate. Subsequently, activated T cells can assist in the activation of B cells, which then produce antigen-specific antibodies (or cytotoxic T cells) which can directly kill infected cells.

This detailed recognition process is the cornerstone of the body's adaptive immune response to pathogens. In the case of B cells, activation results in the cell dividing and differentiating into plasma cells, which are the cells that produce antibodies. For T cells, activation leads to cell division and differentiation into effector T cells, which carry out the functions described above.

Epitalon has been shown to activate the proliferation of lymphocytes in the thymus in older individuals, 30] which could be due to several potential mechanisms.

One possibility is that Epitalon enhances the signaling events that take place within lymphocytes when they encounter an antigen. Another possibility is that Epitalon increases the expression of certain genes that are involved in lymphocyte activation. By promoting lymphocyte activation, Epitalon can potentially enhance the body's natural ability to fight off infections and react to pathogens.

Cytokine Regulation

Cytokines are a diverse group of small proteins that play pivotal roles in cell-to-cell communication and the immune response. They can be produced by a variety of cells, including immune cells like lymphocytes and macrophages and non-immune cells like fibroblasts and endothelial cells. The specific effects of cytokines are wide-ranging, and they depend on the context in which they are released.

In the context of the immune response, cytokines can act as mediators and regulators, coordinating the actions of different immune cells to effectively respond to infections or injuries.

For instance, cytokines like interferons have antiviral properties and can inhibit the replication of viruses. Others, like interleukins, can promote the growth and differentiation of immune cells, enabling a more robust immune response. Cytokines are also important in relation to inflammation which is an integral part of the immune response where tissues respond to harmful stimuli, like pathogens or injuries.

They help to direct immune cells to the site of inflammation, stimulate the repair of damaged tissue, and regulate the inflammatory response to prevent it from causing too much harm.

In essence, the regulation of cytokine production and function is critical for maintaining immune homeostasis and preventing diseases. An overproduction of cytokines can lead to uncontrolled, out-of-control inflammation as with a "Cytokine Storm" as well as autoimmune disorders, while an underproduction can result in impaired immune responses and increased susceptibility to infections.

In a study in 2022, the impact of five peptides, known as Khavinson Peptides®, on inflammation and cell proliferation processes was studied. This includes Epitalon and the study utilized the THP-1 human leukemia monocytic cell line, which is able to transform into macrophages through PMA (Phorbol 12-myristate 13-acetate). This is often used in "in vitro" studies as a potent activator of protein kinase C (PKC) which elicits various cellular responses, such as differentiation, proliferation, and gene expression.

In the context of monocytes, PMA induces differentiation into macrophage-like states, making it a valuable tool for studying the mechanisms of the immune response and the mononuclear phagocyte system in a controlled environment.

Monocytic cells are a type of white blood cells that are crucial components of the immune system.

These cells are part of the mononuclear phagocyte system and are characterized by their bean-shaped nucleus. Monocytes circulate in the bloodstream where they identify and engulf pathogens, dead cells, and cellular debris. Upon entering tissue, they differentiate into macrophages or dendritic cells, which are instrumental in orchestrating an immune response, signaling other immune cells, and repairing tissue damage.

This was done to test the regulatory abilities of these peptides. Notably, the Chonluten tripeptide, obtained from bronchial epithelial cells, reduced monocyte production of tumor necrosis factor (TNF) when the cells encountered the inflammatory stimulus of bacterial lipopolysaccharide in vitro.

The diminished monocyte TNF secretion is associated with an established TNF tolerance response, which lessens inflammatory responses. Moreover, all peptides were found to suppress both TNF and the pro-inflammatory cytokine IL-6 secretion induced by LPS in fully differentiated THP-1 cells.

Additionally, peptide-pre-treated THP1 cells showed less adhesion to a layer of LPS-activated endothelial cells (HUVECs), indicating a lower propensity for inflammation.

This study suggests that these peptides may act as natural assistants in creating TNF tolerance in monocytes and have potential roles as anti-inflammatory agents in macrophages during inflammatory responses and pathogen encounters! 31]

By influencing these interactions, Epitalon can potentially affect the signals that these interactions send within the cells, thereby modulating the immune response.

Alterations in the Expression of Genes Related to Immune Response

Epitalon has also been linked to alterations in the expression of genes related to the immune response. Several studies suggest that Epitalon can stimulate gene expression and protein synthesis during neurogenesis, which could potentially affect the immune response. For instance, one study showed that Epitalon stimulated the expression of Interleukin-2 mRNA synthesis in splenocytes from CBA mice in the absence of specific stimulation. 32]

Interleukin-2 is a cytokine that plays a significant part in regulating immune cells, particularly T cells, which are crucial for immune responses.

Further research indicates that Epitalon not only influences gene expression directly but may also regulate epigenetic mechanisms that control gene expression. Epigenetics refers to changes in gene activity that do not involve alterations to the genetic code but still get passed down to at least one successive generation. One of these mechanisms is DNA methylation, a process that can activate or repress genes.

In the context of aging, it's been observed that immune-related genes show major changes in their expression over the course of normal cognitive aging. By influencing the expression of these genes, Epitalon could potentially counteract some of the changes in immune function that occur with age, enhancing the ability of the immune system to respond to infections and control inflammation.

Rejuvenation of the Thymus Gland

Perhaps most notably, Epitalon has been shown to affect the function of the thymus gland. The thymus gland is a small organ located behind the sternum and in front of the lungs. It's a crucial part of the immune system and is responsible for the development and maturation of T cells.

However, as we age, thymic involution leads to a decrease in the production of T cells and consequently weakens our immune function. Epitalon has been shown to have a significant effect on the function of the thymus gland and is believed to stimulate the rejuvenation of the thymus gland. 33] The thymus gland is most active during early life. It grows until puberty and then gradually begins to shrink throughout adulthood. This shrinking process causes the number of T cells produced by the thymus to also decrease.

Epitalon appears to counteract this process in several ways, although the exact mechanisms are still under investigation. One theory is that Epitalon stimulates the production of certain cytokines, such as Interleukin-7 (IL-7) and Interleukin-22 (IL-22), which have been shown to promote thymus health and function. These cytokines might stimulate thymic epithelial cells, which play a critical role in the maturation of T cells, helping to maintain their population even as we age.

Further Scientific Studies on the Effects of Epitalon on Immune Health

Several studies have investigated how Epitalon may influence the immune system and contribute to overall immune health. The study "Modulating effects of Epithalamin and Epitalon on the functional morphology of the spleen in old pinealectomized rats" by Khavinson et al. (2002) explored the impact of Epitalon on the immune system.

Through immunohistochemical and morphometric analysis, it was found that Epitalon prevented hyperplasia (excessive cell growth) of lymphoid cells in the spleen of pinealectomized rats (rats with their pineal gland removed).

Furthermore, Epitalon also potentiated the decrease in extramedullary hemopoiesis (blood cell production outside the bone marrow). These findings highlight the functional relationship between the pineal gland and the immune system, suggesting that Epitalon's effects on the spleen may be a part of its broader regulatory impact on immune health. 34]

Furthermore, "Antioxidant properties of geroprotective peptides of the pineal gland" by Kozina, Arutjunyan, and Khavinson (2007) examined the antioxidant potential of Epitalon. The findings indicated that Epithalamin, derived from the pineal gland, possesses antioxidant capabilities that can, in certain cases, exceed those of melatonin, another product of the pineal gland known for neutralizing reactive oxygen species (ROS).

In experiments with older rats, Epitalon was shown to not only stimulate the production of melatonin but also exhibit an antioxidant mechanism distinct from that of melatonin. It was discovered that Epitalon could directly exert antioxidant effects and stimulate the expression of antioxidant enzymes such as superoxide dismutase (SOD) and ceruloplasmin. 35] This implies that Epitalon can enhance the body's antioxidant defense system, which may contribute to its geroprotective properties and potential benefits for immune health.

Overall, given the profound impact of epigenetics on our health and state of disease, future research may well reveal novel therapeutic approaches to not only cure disease but altogether prevent degenerative decline.

New formulations and therapies utilizing Epitalon could not only enhance bioavailability and efficacy but also patient success. Strengthened and naturally enhanced immunity could not only prolong vitality and promote longevity, but in the future, it might lead to a lifetime free of illness and disease.

Indeed, the path ahead is filled with promise! As our understanding of Epitalon deepens, we move closer to harnessing its full potential, ultimately contributing to advancements in health and improving human quality of life.

CHAPTER 8

METABOLIC AND ENDOCRINE ENHANCEMENT

One of the biggest challenges many of us face as we age is maintaining our vitality, fitness, and strength. As we age, our bodies naturally begin to lose some degree of muscle mass and experience a reduction in metabolic rate, a process known as sarcopenia. 'Sarcopenia" is a term derived from Greek sarx, meaning "flesh," and penia, meaning "poverty." This "poverty of flesh" is a condition characterized by the loss of skeletal muscle mass and strength, as a direct result of aging.

This progressive decline impacts a person's physical function, can increase the risk of falls, cause overall frailty, and overall quality of life. Out of all the factors of aging, the loss of our physical form is perhaps one of the hardest aspects to confront. We may be just as active, still eat very healthy foods, and inside feel as vital as ever, yet our body appears to lose the youthfulness it once had.

The loss of muscle tissue leads to a decrease in strength and endurance, while the slower metabolism results in an increase in body fat percentage over time. Among all our bodily functions, the metabolic and endocrine system plays a major role in the amount of energy we have available.

Metabolism is a biochemical process that converts the food we eat into cellular energy. This energy is vital for every function our body performs, from physical activities like running and jumping to cellular processes such as repair and growth. Our endocrine system is an integral part of our metabolism and is a network of glands that produce and regulate our hormones.

These hormones act as chemical messengers, controlling a multitude of bodily functions like our mood, cellular growth, reproductive cycles, and the metabolic process itself. Many of us turn to hormone replacement therapies, meal-replacement products, dietary pills (to enhance metabolism or suppress appetite), or work out even harder in an attempt to maintain the younger body we know and identify with. Epitalon has been shown to have a positive impact on our pineal gland and the production of our hormones. It could be a powerful, natural way to maintain our youthful vitality and physical energy as we gracefully age.

How Does Our Metabolism Change As We Age?

Our metabolism is composed of two key processes: catabolism and anabolism. Catabolism is the breaking down of compounds from the food we eat, including the digestion of food and the breakdown of these nutrients into simple molecules for energy. On the other hand, anabolism is the building up of compounds from these nutrients, utilizing energy to create various cells and tissues.

Our metabolic rate, or the speed at which these processes happen, is determined by numerous factors such as our genetics, sex, health status, and age. Additionally, it is influenced by our body's composition. As an example, muscle tissue is metabolically very active and burns more calories than fat tissue does, even when the body is at rest.

Our age significantly impacts our metabolism. As we grow older, our metabolic rate generally slows down. This decrease is primarily due to two factors: a loss of muscle mass and certain hormonal changes that happen naturally but which can also be accelerated or influenced by our environment.

As we age, we naturally lose muscle mass in a process called sarcopenia. Beginning around the age of 30, most people start to lose about 3% to 5% of their muscle mass per decade. As muscles are metabolically more active than fat, this loss leads to a lower metabolic rate, meaning we burn fewer calories at rest than we did when we were younger.

Hormonal Changes

Hormones are chemical substances that our cells produce in the endocrine glands and system. These chemicals act as signaling messengers and, once they are produced, enter the bloodstream where they are transported to various tissues and organs throughout the body. Target cells with specific receptors for these hormones receive them and respond accordingly. This is an essential part of maintaining homeostasis (a balance between several constantly changing factors) and regulating physiological functions. Here, they work to control and manage many bodily functions like cellular growth, our metabolic functions, sexual function, our reproduction, how we feel, our mood, and much more.

Hormones such as insulin, glucagon, and thyroid hormones are important in regulating our metabolic functions. Insulin, produced by the pancreas, facilitates the uptake of glucose by cells for energy which lowers blood sugar levels. Glucagon, on the other hand, works in opposition to insulin and signals the liver to release stored glucose, thus increasing our blood sugar when it is too low. The thyroid hormones, like thyroxine (T4) and triiodothyronine (T3) produced by the thyroid gland, play a significant role by controlling the rate of metabolism, influencing how fast or slow the body consumes energy, and regulating our body temperature. As we age, the levels of certain hormones alter which can affect our metabolic rate. For example, the levels of many thyroid hormones typically decrease as we age, contributing to a slowed metabolic rate.

The Impact of a Slower Metabolism

One of the most notable effects of slowed metabolism is an increase in weight as the body burns calories at a slower pace. This can make weight management quite difficult as the body begins to have trouble losing unwanted weight, compared to how it would otherwise function. Additionally, a slowed metabolic rate is typically followed by reduced energy and fatigue, as the body begins conserving resources instead of "burning fuel." A diminished metabolism can also result in enhanced sensitivity to cold, as less energy used leads to a lowered production of heat. Fluctuations of thyroid hormones can also impact our mood and cognitive function, possibly leading to symptoms of depression or difficulty concentrating.

Aging and Hormonal Imbalances

As we age, the production of certain hormones decreases which can significantly impact our health and quality of life. Some of the major hormones affected by aging include estrogen, testosterone, and insulin.

Estrogen and Aging:

Estrogen is the primary female sex hormone that is responsible for regulating the menstrual cycle and reproductive system. As women approach menopause, usually around their late 40s or early 50s, estrogen levels decrease significantly. This drop in estrogen can lead to various uncomfortable symptoms and changes in the body, such as hot flashes, night sweats, mood swings, and difficulty sleeping. Furthermore, the skin is often one of the most visible indicators of how we age. As we age, it naturally loses a lot of its elasticity and firmness, causing the formation of wrinkles and fine lines.

This process is accelerated by hormonal imbalances, particularly a decline in estrogen levels. Low estrogen levels lead to a decreased production of collagen and elastin, the main proteins that provide the skin with its structure and elasticity.

Estrogen is also essential for the body, in maintaining bone density. Osteoporosis, which is a disease characterized by weakened bones, is more common in older adults and particularly post-menopausal women. Furthermore, estrogen is crucial in how we maintain our balance of "good" and "bad" cholesterol. Estrogen helps to regulate cholesterol by increasing the level of high-density lipoprotein (HDL), or "good cholesterol," and by decreasing low-density lipoprotein (LDL), or "bad cholesterol."

This effect of estrogen on cholesterol levels is particularly noticeable in women during their reproductive years when their estrogen levels are at their peak. Post-menopausal women, however, tend to have lower levels of estrogen and often experience an increase in LDL cholesterol.

Testosterone and Aging:

Testosterone is the primary male sex hormone, and it also decreases as we age. This decline can begin as early as a man's 30s and continue gradually over time. Testosterone is not only the main sex hormone but an essential hormone for overall wellness in men. A decrease in its levels can lead to many uncomfortable physical changes in men, such as reduced muscle mass, low energy levels, lessened stamina, decreased libido, moodiness, fatigue, and symptoms of depression.

Testosterone also affects metabolism as it is used by the body to build muscle mass. As such, a decline in testosterone contributes to a slowed metabolic rate as can often be seen in older men.

Insulin and Aging:

Insulin is a crucial hormone in how our body regulates our blood sugar levels. As we age, our body's response to insulin changes and can lead to increased insulin resistance. This means that the cells of the body become less responsive to insulin. The pancreas either becomes unable to properly produce insulin or the cells become unable to properly respond to insulin, causing heightened blood sugar levels. Over time, this can lead to the development of diagnosed Insulin Resistance, Pre-diabetic symptoms, or Type 2 diabetes.

For the same reasons, the risk of Type 2 diabetes is said to increase with age, yet this risk is mainly observed to be increased with an unhealthy, traditional way of aging.

Cognitive decline is another common sign of degenerative aging, characterized by a gradual decrease in cognitive functions like memory, our attention, and ability to make decisions. Notably, estrogen and testosterone, which are also linked to cognitive functions and neuroprotection, decrease and affect our memory and spatial recognition.

Additionally, levels of dehydroepiandrosterone (DHEA), which is associated with cognitive performance, also decline. The growth hormone (GH) and insulin-like growth factor 1 (IGF-1), both of which are crucial for neuronal growth and survival, similarly diminish. This negatively impacts our brain health and function. Furthermore, the stress hormone cortisol has been shown to cause adverse effects on our memory and accelerate cognitive decline if its levels are consistently elevated over time due to chronic stress. 36] Finally, the hormone melatonin which regulates our Circadian Rhythm and sleep-wake cycle is also seen to diminish with degenerative aging, affecting the quality of our sleep and cognitive abilities.

Current Treatments
for Metabolic and Hormonal Problems

According to the National Institute of Diabetes and Digestive and Kidney Diseases (NIDDK), there is a significant number of endocrine and metabolic-related diseases in US populations. Around 13 million people in the United States, representing 4.78% of the population, suffer from endocrine disorders. 37] Additionally, conditions that have a prevalence estimate of at least 5% in adults include diabetes mellitus and impaired fasting. Metabolic syndrome, which defines a cluster of conditions that increase the risk of heart disease, stroke, and Type 2 diabetes, is also quite common.

Current treatments for metabolic and hormonal problems are quite diverse, as these issues can stem from a variety of conditions. Depending on the specific issue, treatments can range from lifestyle changes to medications, hormone therapy, and even surgery. For hormonal imbalances, hormone replacement therapy (HRT) is typically the primary treatment. This approach involves supplementing the body with synthetic or bioidentical hormones, through injections or pellets, to restore balance. HRT is commonly used for conditions like menopause, hypothyroidism, and low testosterone. Metabolic disorders, on the other hand, often require a more comprehensive approach. Lifestyle changes like diet modifications and regular exercise are typically the first line of treatment. If these measures are not effective, a physician may prescribe specific medications to try and control the blood pressure, cholesterol levels, or blood sugar levels.

Furthermore, certain medications used to medically treat obesity, like gastrointestinal lipase inhibitors and serotonin 2C receptor agonists, come with many side effects.

These include gastrointestinal issues, dry mouth, constipation, and in rare cases, severe liver damage. Aside from this, invasive procedures like stomach reduction surgeries (bariatric surgery) include various procedures designed to alter the digestive system. One widely recognized method is the so-called "gastric sleeve surgery," or sleeve gastrectomy. This involves removing a portion of the stomach to reduce its size, thereby limiting the amount of food one can physically eat, quickly causing a feeling of fullness. This surgery can automatically cause major weight loss fast but also comes with serious side effects like nausea, acid reflux, chronic vomiting, and nutritional deficiencies due to the reduced ability to absorb nutrients.

The Effects
of Epitalon on Our Metabolism

One of the primary roles of Epitalon is its contribution to the regulation of the cell cycle. This means it plays a crucial role in cell division and the growth of new cells, both of which are pivotal processes for maintaining overall health and bodily function. However, recent studies have shown that Epitalon may have a broader range of effects, particularly on our metabolic system.

Epitalon has been shown to rescue the mitochondrial membrane potential during a specific biological cycle, in vitro. 38] Maintaining a normal mitochondrial membrane potential is essential for mitochondrial respiration to occur, which in turn is crucial for the preservation of mitochondrial functionality. The mitochondria control our energy production through a process known as cellular respiration. They convert oxygen and nutrients into adenosine triphosphate, or ATP, the main energy currency of our cells. This process begins in the cytoplasm with glycolysis, which splits glucose into pyruvate.

This produces a small amount of ATP and nicotinamide adenine dinucleotide + hydrogen (NADH). The pyruvate is then transported into the mitochondria, where it enters the citric acid cycle (also known as the Krebs cycle). This cycle further breaks down the pyruvate, releasing CO_2, generating more NADH, and producing another energy carrier, $FADH_2$, as well as a small amount of ATP.

ATP is essential for cellular functions; it provides the necessary energy for a variety of metabolic processes, including muscle contraction, nerve impulse propagation, and chemical synthesis. Without ATP and the mitochondria that produce it, cells would not have the energy required to sustain life! Therefore, it can be said that Epitalon supports this vital function, by way of supporting the mitochondrial membrane potential.

Moreover, Epitalon has been shown to influence other aspects of metabolism. For example, it appears to modulate the function of insulin. A study showed that the administration of Epitalon to aging monkeys reduced the level of glucose and reduced the peak of insulin, 5 minutes after an administration of glucose, compared to the levels of the younger monkeys. Additionally, Epitalon was shown to decrease the area under the plasma glucose response curve, increasing the glucose "disappearance" rate and normalized the plasma insulin dynamics that responded to the administration of glucose. 39] By improving the body's reaction to glucose, Epitalon could potentially help manage conditions related to insulin resistance, such as Type 2 diabetes.

Furthermore, there is evidence suggesting that Epitalon may have beneficial effects on lipid peroxidation through oral ingestion in fruit flies. 40] Lipid peroxidation has been linked to human genetic diseases and cancer, so it is interesting to consider what other beneficial reactions Epitalon may produce.

The Influence
of Epitalon on Hormonal Health

Epitalon has also been shown to have a profound influence on hormonal health. One of the primary ways it does this is by interacting with various aspects of the endocrine system, which plays a crucial role in regulating hormone production and function. One of the key hormones influenced by Epitalon is melatonin, which is produced by the pineal gland in the brain. It plays a crucial role in regulating sleep-wake cycles, also known as circadian rhythms. Production of this hormone increases in the absence of light and decreases with light exposure, hence why it is often referred to as the "sleep hormone" or "darkness hormone." In addition to its primary function in sleep regulation, melatonin also possesses antioxidant properties. Its ability to neutralize harmful free radicals and reduce oxidative stress contributes to its role in maintaining overall cellular health.

Studies have shown that Epitalon increased the basal night melatonin in aged rhesus monkeys, indicating its potential to positively affect the pineal gland and normalize the production of insulin. 4] This could help improve the quality of sleep and the length of time we are able to stay asleep.

Furthermore, Epitalon has also been shown to influence the hypothalamus. This is the center of the brain that is part of controlling the body temperature, hunger, and mood. Epitalon seems to enhance the sensitivity of the hypothalamus and affect the hypothalamic-pituitary-gonadal axis. 41] By doing so, Epitalon could potentially help balance the body's energy usage and how it maintains a healthy weight.

Research on Epitalon
and Age-Related Metabolic Changes

Certain studies and research on how Epitalon influences metabolic markers have explored its effects. These studies offer valuable insight into the potential role of Epitalon in modulating certain metabolic pathways.

Epitalon and Thyroid Gland Structure:

A study conducted by Kuznik et al. (2011) investigated the impact of two peptides: Lys-Glu-Asp-Gly and Ala-Glu-Asp-Gly (Epitalon. Their effect on the hormonal activity and thyroid gland structure in hypophysectomized (removal of the pituitary gland) young chickens and old hens was examined. In the context of Epitalon (Ala-Glu-Asp-Gly), the researchers observed noteworthy effects on hormonal imbalances and structural changes in the thyroid gland, induced by hypophysectomy.

After subjecting the birds to a 40-day treatment with the synthesized peptides, derived from the amino acid composition of the anterior and posterior lobes of the pituitary gland, a significant reduction in the observed changes was noted.

The peptides, including Epitalon, demonstrated a normalizing effect on the hormonal activity and structural integrity of the thyroid gland. This suggests a potential regulatory role of these peptides in mitigating the impact of hypophysectomy-induced disruptions.

However, the study also reported that while the normalizing effects were evident in both young chickens and old hens, the impact on thyrotrophic hormone and thyroid hormone concentrations was less pronounced in the older subjects compared to the younger ones. This distinction highlights a potential age-related variability in the response to Epitalon treatment.

The findings from this study suggest that Epitalon, along with the peptide Lys-Glu-Asp-Gly, may have a beneficial impact on hormonal activity and thyroid gland, particularly under the conditions of hormonal imbalances induced by hypophysectomy. 42]

Epitalon and Reproduction:

In another study, conducted by Korenevsky et al. (2013), the effects of melatonin and Epitalon on the central regulation of reproduction in female rats exposed to unfavorable environmental factors were examined. The focus was on understanding the neuroprotective aspects of these substances, particularly in the context of the hypothalamic regulation of hormonal activity and metabolism.

The research involved examining estrous cycles in rats of different age groups (young, mature, and aging) subjected to light pollution. The study assessed the diurnal dynamics and daily mean content of biogenic amines in hypothalamic areas responsible for gonadotropin-releasing hormone (GnRH) synthesis and secretion.

The results indicated that the administration of melatonin and Epitalon played a corrective role in addressing impairments in the hypothalamic-pituitary-gonadal axis caused by continuous artificial lighting and exposure to the neurotoxic xenobiotic 1,2-dimethylhydrazine. These substances were found to have a protective effect on the reproductive system of both young and aging female rats, especially in the face of unfavorable environmental influences.

The study underscored the significance of the pineal gland in forming the circadian signal crucial for the preovulatory peak secretion of GnRH. Melatonin and Epitalon were suggested to mitigate the adverse effects of environmental stressors on the reproductive function of female rats across different life stages. 43]

The Benefits Of Epitalon
For Weight Management

Managing our weight can sometimes seem like a complex process where the body does not always respond as we would like it to. Primarily, issues with losing or gaining weight stem from problems with the regulation of our metabolism, appetite control, or our hormonal balance. Each of these plays a significant role in maintaining a healthy weight and overall wellness.

Metabolism Regulation:

The metabolic rate is influenced by factors like our age, gender, the muscle-to-fat ratio, and level of physical activity, as previously mentioned. As opposed to merely relying on fad diets, dietary pills, or meal-replacement products, regulating our metabolism naturally can be very beneficial. Aside from a healthy diet and regular exercise, Epitalon may help the body maintain a healthy metabolic rate, to more effectively burn calories and prevent excess fat from being stored.

Appetite Control:

Appetite control is another vital aspect of weight management. The brain, specifically the hypothalamus, signals the body whether we are hungry or full. Hormones such as ghrelin, also known as the "hunger hormone," stimulate our appetite, while the hormone leptin, produced by fat cells, sends signals to the brain to suppress appetite when sufficient energy has been stored.

Maintaining a balanced diet and regular eating patterns can aid in proper appetite control. Additionally, Epitalon may be able to support the hypothalamus and promote a normalization of its function.

Promoting Protein Synthesis:

In addition to enhancing mitochondrial function, Epitalon may boost energy levels by promoting protein synthesis. 44] Proteins are fundamental to virtually every biological process, including energy production. They serve as the building blocks for enzymes, many of which are involved in metabolic processes that facilitate the production of energy. Epitalon's ability to stimulate protein synthesis could thus have far-reaching implications for metabolic health.

By increasing the production of these energy-related enzymes, Epitalon can potentially enhance metabolic processes. This could lead to an increase in energy production and availability, further contributing to overall metabolic efficiency and health. The implications of this are vast, ranging from improved physical performance to enhanced recovery and healing processes. Moreover, by promoting protein synthesis, Epitalon may also contribute to muscle growth and repair, another critical aspect of metabolic health.

The ability to efficiently build and repair muscle tissue is crucial for maintaining metabolic health, as muscle is a metabolically active tissue that contributes significantly to energy expenditure.

The Future of Epitalon
for Metabolic and Endocrine Enhancement

Epitalon's influence on the cardiovascular, endocrine, and immune systems suggests it could be pivotal in the advancement of various therapies and anti-aging therapies. One of the intriguing aspects of Epitalon is its suggested ability to enhance the activity of telomerase, a vital enzyme not only involved in cellular aging but also in lipid metabolism.

By modulating these processes, Epitalon could positively impact our energy production, the way our body stores energy, and the overall regulation of our metabolism. In the future, we may see tailored therapies designed for people who are otherwise genetically predisposed to metabolic diseases, or who have inherited tendencies to be overweight or develop thyroid issues.

As Epitalon also affects our neuroendocrine system and the pineal gland, this short peptide can stimulate our gene expression and regulate the function of vital organs like the retina and brain. Therefore, we might very well see Epitalon utilized in the management of traditional age-related disorders in the field of longevity and enhanced quality of life.

In terms of complementary therapies, Epitalon could be integrated into a broader treatment plan aimed at enhancing metabolic and endocrine functions. Beneficial lifestyle modifications, like a balanced diet and regular exercise, are well known to improve metabolic health and could be combined with Epitalon therapies for a more comprehensive approach. Epitalon could possibly enhance these naturally beneficial practices, by supporting how the mitochondria produce and handle cellular energy.

Overall, Epitalon may in the future be seen as a healthy and natural way to manage our weight and maintain a healthy body, which is increasingly challenging as we age. It may in the future replace artificial hormones and entirely reduce the need for excessive supplements, maintaining youthful energy well into our golden years.

CHAPTER 9

GOOD QUALITY SLEEP FOR IMPROVED LONGEVITY

In today's dynamic, fast-paced society, we often find ourselves sacrificing our sleep for productivity and obligations. The constant connection to work and the relentless hustle and bustle of life frequently result in our sleep being compromised and shortened. This has led to a widespread issue of chronic sleep deprivation with profound implications for our health, welfare, and our lifespan. Sleep is not merely a time of rest—it is a fundamental aspect of our overall wellness. It is the foundation for several critical functions, like memory consolidation, detoxification of the brain and body, regulating our metabolic functions, our hormonal balance, and restoring the body overnight.

Despite its importance, it seems that a significant proportion of the global population is not receiving adequate sleep. According to statistics, the majority of adults typically achieve between 6 to 8 hours of sleep every night, with 27.5% managing to get exactly 7 hours. While the specific hours may vary, the average sleep duration for both genders is approximately 6.73 hours per night. When it comes to getting a full 8 hours or more of sleep, about 33% of men manage to do so. This is slightly more than the 31% of women who achieve the same. 45]

This is a global issue and, alone is problematic, but an even more concerning scenario is developing here in the United States. The Centers for Disease Control and Prevention states that around 1/3 of adults in the US do not get sufficient rest or sleep daily.

Almost 40% of these adults unintentionally fall asleep during the day at least once a month. Moreover, it is estimated that between 50 to 70 million Americans suffer from chronic sleep disorders. 46] This widespread shortage of sleep is caused by many factors. The increase in technology has led to us spending more time staring at screens and working longer hours, causing a lot of the growing stress levels we now see. Plus, our culture tends to push a sense of non-stop busyness while ignoring the need to rest, which worsens the problem. We need to shift our perspective on the importance and value of sleep. It is high time we recognize sleep for what it truly is—an indispensable part of maintaining our health and enriching our lives!

Good Quality Sleep and Anti-Aging

The importance of sleep in relation to anti-aging is often underestimated, but its role is fundamental. Sleep is a crucial factor in aging where the body is provided with the time it needs to properly heal and "reset" itself. While we sleep, our bodies are busy restoring and repairing any damage from stress we may have experienced from UV light and other harmful influences during the day. This restoration process involves an increase in certain proteins by our cells to facilitate necessary cellular repair.

One of the most notable aspects of sleep is its relation to the body's release of growth hormone during the deep sleep stages. This hormone, also known as somatotropin, is a protein hormone that plays a crucial role in human development. It stimulates growth, cell reproduction, and cell regeneration. The role of this hormone extends to maintaining the firmness and elasticity of our skin by promoting the production of collagen, which is a vital protein that gives our skin structure and strength. As we age, our collagen production naturally slows, leading to sagging, fine lines, and wrinkles. However, the growth hormone our body releases during deep sleep can boost the production of collagen, which can help delay these visual signs of aging.

Beyond our skin's health, the quality of sleep is essential to the overall health and longevity of our body's various systems. During sleep, our body undergoes numerous repair processes, including those related to our cardiovascular system. The heart and blood vessels are repaired and restored during sleep, and chronic sleep deprivation has been linked to increased risk of cardiovascular disease and hypertension.

Sleep also significantly impacts our metabolic system. It is important in maintaining a healthy balance of the hormones which regulate our feelings of hunger and satiety. Ghrelin, for instance, which is often referred to as the "hunger hormone," sends signals to our brain when it is time to eat. Leptin, on the other hand, or the "satiety hormone," signals our brain to indicate that we are full and should stop eating. However, a lack of sleep can disrupt the balance of these two hormones and override other signals. This can cause us to sense that we are hungry, although we are full, and leads to overeating, unhealthy snacking, increased weight gain, and its associated health risks.

Furthermore, sleep also plays a central role in our cognitive function and the health of our brain. When we sleep, the brain goes through a "clean-up" process where harmful toxins, accumulated throughout the day, are cleared out. An article from Yale's School of Medicine has commented that two vital processes occur during the sleep cycle: Memory Encoding and Memory Consolidation. Memory encoding is the initial process where the brain captures and interprets stimuli from the external environment, creating networks of neurons in the hippocampus that sequence the events or experiences. These neural sequences represent the intricate details of the memory, with the amygdala (a specific part of the brain) acting much like a filter, intensifying the storage of memories linked to emotional events. This is why we often remember emotionally charged moments more vividly than neutral ones. It also is thought to add emotional relevance to them as they are formed.

On the other hand, memory consolidation is the phase that follows, where these encoded memories are strengthened and integrated into long-term storage during sleep, predominantly during the "slow-wave phase." This integration links the newly encoded sequences with pre-existing knowledge networks within the neocortex. The neocortex is a part of the brain in mammals that's involved in higher-order functions such as sensory perception, cognition, generation of motor commands, spatial reasoning, and language. Without consolidation, even well-encoded memories may fail to be retained, emphasizing why sleep is so crucial for the preservation of episodic memories. 53]

Poor sleep, interrupted sleep patterns, or a lack of sleep can all impair these cognitive functions and lead to problems with attention span, decision-making, ability to concentrate, and our mood.

How Poor Sleep Affects Aging

As we age, our sleep patterns naturally change, but consistently poor sleep or sleep deprivation can accelerate the aging process. When we don't get enough good quality sleep, our body misses out on the crucial restorative processes that happen during sleep.

One of the most visible results of poor sleep is seen on our skin. Chronic sleep deprivation leads to an increase in the production of cortisol, a well-known stress hormone. High levels of cortisol can break down collagen, which is the protein that provides structure and elasticity to our skin. As collagen breaks down, our skin begins to lose its firmness and elasticity, resulting in wrinkles, sagging, and other visible signs of aging. A study has shown that there is a clear, visible distinction in those who get good quality sleep and those of us that do not. It was shown that poor sleepers have a significantly higher amount of trans-epidermal water loss, or dehydration of the skin. 54]

In addition to the visible signs of aging, poor sleep can accelerate aging internally as well. Numerous studies have found a link between insufficient sleep and a range of health issues such as heart disease, high blood pressure, obesity, and diabetes. These conditions can not only speed up the aging process but also significantly reduce the quality of life.

The Importance of Quality Sleep for Longevity

Numerous scientific research studies have shown that good quality sleep plays a critical role in physical health, mental well-being, and overall quality of life. In one study, the relationship between the duration of sleep, sleep disturbances, and life expectancy (LE) without chronic disease, in individuals aged 50 to 75, was examined. Data collected from three occupational cohort studies in England, Finland, and Sweden were used, with a follow-up period ranging from 6 to 18 years.

The study found that those who slept for 7 to 8.5 hours and reported no sleep disturbances had the longest, healthiest, and chronic disease-free LE, living approximately 19.1 years in good health and 13.5 years without chronic diseases, from the age of 50 to 75. Those who slept less than 7 hours, however, or more than 9 hours, or reported severe sleep disturbances, had a 1-3 years shorter healthy LE and a 3 to 6 years shorter chronic disease-free LE. 47]

In another study, researchers sought to analyze sleeping patterns and biochemical profiles of individuals at the age of 85-105 years old, comparing these to young adults (20 to 30 years old) and older adults (60-70 years old). The research involved polysomnography (a comprehensive sleep disorder test, used as a diagnostic), a week of monitoring their human rest/activity cycles, and blood samples. The results showed that the oldest individuals had a poorer level of quality sleep and REM sleep compared to the other adults.

These older individuals had strict, regular sleep-wake schedules and presented higher HDL-cholesterol and lower triglyceride levels than the other adults. 48]

Furthermore, another interesting study was done by Hou et al. (2020), which evaluated the impact of the duration of sleep on all-cause mortality and quality of life in individuals aged 80 and above. Data from 15,048 participants were analyzed.

The results revealed a U-shaped association between the duration of sleep and all-cause mortality, with the lowest risk of death identified at around 8 hours of sleep. A J-shaped association was found between the sleep duration and a poor quality of life. Compared to a time of sleep of about 7 to 9 hours, both shorter periods (less than 7 hours) and longer periods (more than 9 hours) of sleep were associated with an increased risk of death. 49] This study shows that there is a correlation between how well we sleep and the quality of life overall, especially in our later years.

Epitalon and Improvement of Sleep Patterns

One of Epitalon's most notable effects is seen in its potential to regulate the circadian rhythm. This is the "internal body clock" we all have which governs various biological processes, including our sleep-wake cycle. The circadian rhythm is a natural, internal system designed to regulate feelings of sleepiness and wakefulness over a 24-hour period. This rhythm can be disrupted by factors such as irregular sleeping schedules, exposure to artificial light at night, or aging. When this happens, it can lead to a host of issues, from insomnia to chronic fatigue.

Epitalon's influence on the pineal gland and, consequently, on the circadian rhythm extends beyond just regulating our sleep. A primary role of the circadian rhythm is in regulating the release of certain hormones. Hormones are the body's chemical messengers and they control everything from our growth and development to our metabolism and mood. As an example, cortisol typically peaks in the morning to help us wake up and decreases throughout the day. An imbalance of this rhythm can lead to chronic stress, fatigue, and many other health issues.

By stimulating the pineal gland and enhancing melatonin production, Epitalon may help reset the circadian rhythm and ensure optimal timing of the release of hormones to promote a better hormonal balance.

The circadian rhythm also plays a crucial role in regulating our eating habits and digestion. Our bodies are designed to eat during the day and rest at night. Disrupting this rhythm, like eating late at night, ultimately leads to metabolic disorders like obesity and Type 2 diabetes. By regulating the circadian rhythm, Epitalon can potentially help normalize eating patterns and improve metabolic function, which contributes to a healthier body weight and a reduced risk of metabolic diseases. Our body temperature is another key physiological process that is controlled by the circadian rhythm. The body temperature fluctuates throughout the day, is usually lowest in the early morning, and highest in the late afternoon. This rhythm is crucial for various bodily functions, such as metabolism and sleep. Disturbances in this rhythm can also lead to sleep disorders and impaired physical performance.

Epitalon's potential to regulate the circadian rhythm could help maintain our body temperature naturally, contributing to better sleep and optimal physical functioning. As the circadian rhythm also influences many other important bodily functions, like the regulation of our blood pressure, kidney function, and immune response, supporting a healthy circadian rhythm, Epitalon could support these vital processes and promote physiological harmony. This could potentially reduce the risk of various health issues like hypertension, kidney disease, and immune-related disorders.

Epitalon's Effect on Melatonin

Melatonin is a hormone that is synthesized and released by the pineal gland in response to darkness, signaling our body when it's time to sleep. As we age, the production of this hormone decreases due to various factors, which can lead to problems such as insomnia or fragmented sleep. Research indicates that Epitalon may stimulate the pineal gland, enhancing its ability to produce melatonin. This stimulation can help recalibrate our internal biological clock, helping to regulate sleep patterns and ensure they align with a natural day-night cycle. This can help us fall asleep quicker, experience fewer wakings during the night, and result in more restorative, deep sleep.

In one study done by Arutjunyan and colleagues, it was found that melatonin and the pineal gland peptides (Epithalamin, the natural form produced by the pineal gland, and Epitalon, the synthetic form) can correct the disturbed diurnal dynamics of norepinephrine (NE) in the medial preoptic area (MPA) and dopamine (DA) in the median eminence, with arcuate nuclei (ME-Arc).

These disturbances were caused by a single administration of the neurotoxic xenobiotic 1,2-dimethylhydrazine (DMH) in female rats. The study indicated that experiments with administering DMH could serve as an animal model of premature aging in the female reproductive system. It showed that Epithalamin could restore the disrupted circadian rhythms caused by DMH.

Furthermore, Epitalon has been shown to be more effective in preventing the disturbance of DA diurnal rhythm caused by the xenobiotic, maintaining low levels at 5 o'clock Circadian Time (CT) and increasing it by 11 o'clock (CT).

By demonstrating the ability of melatonin and these pineal peptides to correct the diurnal dynamics of certain neurotransmitters, this suggests potential pathways through which these could help regulate our sleep patterns and overall quality of sleep. 50]

By helping to improve sleeping patterns, Epitalon could support our cognitive function and emotional health and contribute to a more stable, positive mood. During sleep, the immune system also releases specific proteins called cytokines, which help to promote sleep and combat infection or inflammation. As sleep is also essential for our immune system, improved sleep, facilitated by Epitalon, could lead to a stronger immune response, increasing our resistance to diseases and infections.

Scientific Studies on Epitalon and Improved Sleep

Numerous studies suggest that Epitalon may contribute to more regular circadian cycles and improved sleep quality. The research study titled "Synthetic tetrapeptide epitalon restores disturbed neuroendocrine regulation in senescent monkeys," conducted by V Khavinson, N Goncharova, and B Lapin, focused on the impacts of Epitalon on aging monkeys. The study utilized female Macaca mulatta (rhesus) monkeys of varying ages. During the study, levels of melatonin and cortisol were assessed with an immunoassay.

The findings indicated that administering Epitalon significantly stimulated the synthesis of melatonin in aging monkeys during the evening hours. This is notable as it led to the normalization of the circadian rhythm of cortisol secretion, which is crucial for maintaining various physiological processes. Therefore, Epitalon could potentially play a key role in supporting neuroendocrine regulation in senescence. 51] In another study, researchers found that when Epitalon was administered to the older monkeys, it led to increased basal night levels of melatonin. 52] This research indicates that Epitalon has a positive effect on normalizing melatonin levels and that it may also restore age-related endocrine dysfunctions in primates.

Combining Epitalon
with Other Lifestyle Changes

Regardless of what approach or treatment we employ, incorporating better sleeping habits into your daily routine greatly affects the outcome. Firstly, establishing a regular sleep schedule is crucial. The body operates on a natural, internal clock, which is significantly influenced by exposure to light. Especially the artificial "blue light" from our smartphones, TVs, or other devices have been shown to disrupt our natural clock.

By going to bed and waking up at the same time every day, you can help synchronize your body's internal clock with the external day-night cycle. This consistency can lead to improved sleep quality, as your body gets accustomed to a set sleep pattern. Secondly, creating a restful and relaxing environment can facilitate better sleep. An ideal room to sleep in is usually cool, dark, and quiet. This can be created by using darkening curtains, earplugs, fans, or white noise machines. Creating an actual bedtime ritual can also be extremely beneficial.

Much like children have routines that prepare them for bed, adults can also benefit from relaxing pre-sleep rituals! For instance, reading a relaxing book, taking a nice warm bath, listening to some calming music, or doing some simple stretches can help prepare you for sleep by relaxing your mind and body. This signals your body that it's now time to "wind down" and get ready for some good sleep.

Furthermore, what you eat and drink during the day can also affect how well you sleep. Caffeine should naturally be limited and keep in mind that some of us are more sensitive than others - depending on our genetic variants. Also, avoid eating any large meals close to bedtime. Always allow at least three hours after you finish eating BEFORE you go to bed!

This is important because your body needs to go into detox mode and cannot do that if digestion is occurring.

Caffeine in your body or having to digest food can disturb your sleep cycle and can lead to poor quality of sleep. On the other hand, regular physical activity during the day is another important aspect of good sleep. Regular exercise can help you fall asleep faster and enjoy a deeper sense of sleep as the body is well spent and naturally exhausted. However, the timing is important—exercising too close to your bedtime can have the opposite effect as it can stimulate the body and make it harder to fall asleep.

Potential Advancements
in Epitalon Research for Sleep and Longevity

As science continues to discover the complexities and benefits of Epitalon, new implications for longevity are bound to be uncovered! Further understanding what pathways Epitalon influences and how this can improve our sleep on a cellular level could lead to new therapeutic treatments for sleep disorders. Given the critical role of sleep in overall health and well-being, this could have far-reaching implications.

This might prevent many of the current sleep disorders from developing or offer an actual treatment for the ones that at this time have none. Additionally, the research into Epitalon's effect on our endocrine system and the pineal gland could result in new data on how we sleep and what affects our quality of sleep. Adequate and proper sleep-wake cycles are essential for our rest, the body's rejuvenation, and detoxification. It is also crucial in reinforcing the immune system, managing our metabolism, and preserving our cognitive functions—all of which serve as preventative measures against chronic illnesses and premature aging. By enhancing the natural way we sleep, Epitalon could reduce any dependence on conventional sleep aids which often come with dangerous side effects and the risk of addiction.

It is known in the industry of QEEG and neurofeedback (my area of Board Certification) that those sleep aids only bring on sleep fifteen minutes earlier than a person would without them! They don't help people sleep better! They make people forget how poorly they slept! By promoting a more natural approach, Epitalon aligns with the rising preference for treatments that support the body's intrinsic systems with minimal disruption to its biological processes. As research progresses, we can look forward to a deeper understanding of Epitalon and its potential applications in promoting longevity and enhancing our quality of life!

CHAPTER 10

SKIN HEALTH AND EPITALON

Although many of us take great pride in our skin, we tend to overlook its role in our health and its connection to our overall well-being. Aside from the aesthetic value of the skin, which often becomes the general focus when we wish to take care of it, our skin is a key indicator of our health and an essential part of our immune system.

Our skin is the largest organ of our body and a crucial part of our immune system. It serves as our primary defense against harmful environmental factors, such as bacteria and toxins, regulates our body's hydration level, and is one of the body's main detoxification pathways.

However, current statistics suggest that one in five Americans will face a skin cancer diagnosis in their lifetime. Currently, approximately 9,500 people in the U.S. receive this life-altering news daily. A vast number of these diagnoses are composed of nonmelanoma skin cancers, such as basal cell carcinoma and squamous cell carcinoma, which affect over 3 million Americans annually. 55] Right now, skin diseases globally account for 1.79% of the overall disease burden, as measured in disability-adjusted life years or "DALYs." In 2019, about 4.9 billion new cases were reported worldwide, indicating that skin conditions affect an estimated 1.8 billion people at any time. The Global Burden of Disease Study 2019 analysis showed that most new cases were fungal and bacterial skin diseases, resulting in roughly 98,522 deaths. 56]

The beauty and media industries often showcase flawless skin as the main symbol of true beauty and health. "Pure skin" has long been seen as a sign of vitality, youthfulness, and attractiveness. Yet, one of the most classic signs of degenerative aging is seen in fine lines, loss of suppleness, wrinkles, or a dry and dull complexion. However, although our production of collagen diminishes as we age, this is not an automatic or natural development of the aging process. How we age has a lot more to do with how we look than the actual number of years in our age!

Our skin is a complex organic structure made up of two primary layers: the outer layer, known as the epidermis, acts as a protective barrier. The layer underneath is called the dermis. This is a collagen-rich layer that offers support and nourishment. A weakened, dry, and deficient skin in hydration and nourishment leaves us with a diminished ability to detoxify and protect ourselves from bacteria, aside from the loss of aesthetic value.

The Most Common Signs of Aging on the Skin

As we age, the changes in our skin become more evident and permanent. The most common signs of aging skin are wrinkles, fine lines, age spots, and less supple skin texture (dry or sagging).

Wrinkles initially appear as lines or creases on the skin, predominantly on the face. They are also found in areas like the hands, neck, and arms. These range from mere surface lines to deeper furrows that form in different areas of the face. For example, "expression lines" typically develop around the eyes and the mouth where the skin moves the most, caused by the flexing of facial muscles and the muscles of mastication. Other types of wrinkles may appear across the forehead or at the outer corners of the eyes, commonly known as "crow's feet."

Sagging skin is caused by the loosening or "drooping" of the skin and a loss of collagen and moisture. It is most noticeable on the face, neck, and hands.

Over time, the skin appears less firm and less tight than it used to be when we were younger. On the face, this can lead to the formation of jowls or droopy eyelids. The skin can also begin to sag in areas like behind the arms, on the breasts, and the abdomen.

Additionally, liver spots (or age spots) and solar lentigines are small, flat, dark areas that vary in size but are usually larger than freckles. Age spots are often tan, brown, or black and are commonly found in areas most exposed to the sun, like the face, hands, shoulders, or arms. They can also appear on the skin of young people, especially those who commonly expose themselves to sunbathing or sun tanning. Despite their name, these skin spots have no relation to our liver function or liver health. They result from the skin's melanocytes, or pigment-producing cells, overproducing melanin in response to prolonged exposure to ultraviolet light. This results in the appearance of these blemish-like concentrated patches of pigment on the skin known as "liver spots" or "age spots."

The Impact Of Aging
On The Skin's Elasticity and Vitality

The aging process affects our skin in several ways, leading to a loss of elasticity and vitality. Underlying physiological processes like collagen degradation, decreased production of elastin, and reduced skin cell turnover all contribute to these changes.

Collagen Degradation

Collagen is the most abundant protein in our body, making up about 75-80% of our skin. It serves as a key building block for skin and provides structure and firmness. However, the production of collagen slows down as we age, and the existing collagen we store in our bodies can become damaged through a process known as collagen degradation.

Collagen degradation is primarily driven by enzymes called matrix metalloproteinases (MMPs) that break down collagen and other components of the extracellular matrix. This matrix is the network of proteins and polysaccharides that provide structural and biochemical support to cells. Increased activity of MMPs due to factors like UV radiation and inflammation can lead to an imbalance between the breakdown and synthesis of collagen, resulting in a net loss of this vital protein. As a result, it weakens the skin's structural integrity and causes it to lose its firmness, leading to the formation of wrinkles and lines.

Decreased Elastin Production

Elastin is another key protein that gives our skin its elasticity, or the ability to stretch and snap back into place. The production of elastin likewise decreases with age, and the existing elastin becomes fragmented and less functional. As a result, this affects the skin's ability to return to its original shape after being stretched or contracted. Every day, we make countless facial expressions as we communicate and express our emotions. When our skin loses its elasticity, the result is seen in the formation of "permanent lines" or "wrinkles" where our facial skin is flexed or moved the most.

Reduced Skin Cell Turnover

Skin cell turnover is the process by which new skin cells are developed for the skin's surface to replace old, dead skin cells that have been shed from the surface. In young, healthy skin, this process takes about 28 days. However, it slows as we age, meaning that dead skin cells remain on the surface of the skin longer. This is part of what can make the skin look dull, colorless, and dry. Moreover, the slowed turnover also means that the skin starts to take longer to heal and repair from damage or injury.

External Factors
that Accelerate Visible Signs of Aging

External factors also contribute to the acceleration of the visible signs of aging, known as "extrinsic aging." Prolonged sun exposure is one of the worst and leading causes of premature skin aging. Also called "photoaging," this relates to the damaging effects of ultraviolet (UV) radiation on the skin's surface, penetrating its layers and causing damage to the collagen and elastin. UV radiation stimulates the production of Matrix Metalloproteinases, or "MMPs" (mentioned above), leading to the degradation of collagen and elastin.. These enzymes play a pivotal role in skin aging and damage because of prolonged UV exposure. They are activated by UV radiation and lead to the breakdown of collagen and elastin, which are critical for maintaining the skin's firmness and elasticity.

This degradation process in turn contributes to the development of wrinkles, a loss of skin tone, and other signs of photoaging. This also results in a loss of elasticity, further leading to wrinkles and sagging. Moreover, UV radiation from sun rays or tanning also increases the production of melanin, which is the pigment that gives our skin its color. Although a darkened skin tone is considered aesthetic or attractive, this can lead to the formation of age spots.

Furthermore, environmental pollutants such as particulate matters, gases, heavy metals, and chemicals also accelerate the aging of our skin. These pollutants generate reactive oxygen species that cause oxidative stress in the skin. As a result, our skin cells and extracellular matrix get more damaged and inflamed.

Some pollutants, like polycyclic aromatic hydrocarbons, or PAHs, can directly bind to and damage collagen and elastin and eventually lead to their degradation. Polycyclic Aromatic Hydrocarbons are a group of organic compounds composed of multiple aromatic rings, made up of carbon and hydrogen atoms.

They are primarily formed through the incomplete combustion of organic materials like coal, oil, gas, wood, and tobacco from cigarette smoke. PAHs are widespread in our environment and are well-known for their potential to pose significant health risks to humans and wildlife, including carcinogenicity (the ability to cause cancer) and mutagenicity, depending on their specific structures and the exposure levels. Some PAHs have also been known to be activated by UV exposure, further increasing their harmful effect. Long-term exposure to these pollutants can result in visible signs of aging, such as wrinkles, fine lines, and changes in skin tone and texture. 57]

Smoking is one external factor that contributes to the premature or accelerated aging of the skin. Cigarette smoke contains thousands of chemicals, many of which can harm the skin. For instance, nicotine is an excellent vasoconstrictor, meaning that it narrows the blood vessels in the outermost layers of the skin and limits the supply of oxygen and nutrients to the skin. Smoking also generates a large amount of free radicals and causes oxidative stress in the skin, in addition to the harmful chemicals that enter through the lungs to the rest of the body. The result of this is an accelerated degradation of our collagen and a decreased production of elastin.

The Influence of Genetics on the Aging of Our Skin

Our genes also influence how our skin ages by affecting collagen production and how well our skin can repair itself. The rate at which our bodies produce and degrade collagen can significantly influence the aging process, and these rates are largely determined by our genes. Our genes code for the proteins that make up collagen and the enzymes that regulate its production and breakdown. However, there are specific genetic variations or even mutations that can lead to differences in the structure and function of collagen.

For instance, certain genetic variants can result in increased activity of MMPs. Like what was mentioned earlier, this increased activity can lead to a "net loss" of collagen, resulting in skin that is more prone to wrinkling and sagging.

Our body is naturally equipped with natural repair mechanisms that help manage damage to the skin, specifically those inflicted on collagen and elastin caused by factors like UV radiation and environmental pollutants. These mechanisms involve a complex network of proteins, all of which are encoded by our genome. For example, our genes code for antioxidant enzymes that neutralize harmful ROS produced by UV radiation and pollution. Other genes are involved in DNA repair to fix mutations caused by said radiation. On the other hand, some people with certain genetic variants may have a less effective repair mechanism, making their skin more prone to damage and signs of aging. For example, some variations in the genes coding for antioxidant enzymes can lead to a decreased ability to neutralize ROS. Similarly, variations in DNA repair genes can lead to an increased risk of mutations, accelerating skin aging.

The Importance
of Skin Care Routines and Lifestyle Choices

While we cannot alter our inherent genetic makeup, we can certainly influence how it manifests through how we care for our skin and health, as well as our lifestyle choices. These factors can help counteract the effects of genetic and environmental influences on the aging of our skin. A well-formulated skincare routine can help protect the skin from damage, promote cellular repair, and stimulate collagen production, which can help slow down the aging process.

For instance, avoiding excessive and harmful exposure to UV radiation from the sun (or sun tanning) may prevent photoaging and help maintain healthy and vital skin in one's later years.

Topical antioxidants, like vitamin C, vitamin E, and resveratrol, help to neutralize harmful ROS to prevent oxidative stress and inflammation. Retinoids, which are derivatives of vitamin A, can also stimulate collagen production and cellular turnover. As such, they help to reduce wrinkles and improve skin texture. Additionally, there are many pure, non-chemical-based moisturizers that can help maintain the skin's barrier function, prevent water loss, and keep the skin hydrated and supple. Incorporating regular exfoliation into a skincare routine can also help to remove dead skin cells and promote cellular turnover to reveal fresher, younger-looking skin.

Furthermore, it goes without saying that healthy lifestyle choices, such as avoiding smoking and drinking enough water, help manage the effects of genetic predispositions and environmental factors. A balanced diet rich in high-quality fruits, vegetables, lean proteins, and healthy fats can provide the nutrients necessary for good skin health, including antioxidants, vitamins, and minerals. In addition, any kind of regular physical activity that promotes circulation can help improve the delivery of oxygen and nutrients to the skin to keep it healthy.

How Epitalon Promotes Skin Elasticity, Firmness and Resilience

Many new technologies and advancements are currently being developed within the field of anti-aging and longevity in the pursuit of youthfulness and vitality. Novel therapies for aging find ways to harness the genetic code of the body and are now better able to understand and influence the aging process itself. However, we are seeing an increase in the desire for regenerative and holistic methods that exploit and boost the natural capabilities our bodies already have.

Epitalon and Its Mechanisms of Action

Aside from the many other benefits covered in this book, Epitalon has been shown to stimulate the synthesis of collagen and elastin by upregulating the expression of specific genes that code for these proteins. This increased production enhances the skin's elasticity, firmness, and resilience and contributes to a more youthful appearance.

Peptides, especially short peptides consisting of 2-7 amino acid residues, have been shown to hold amazing potential in the realm of gene regulation and therapeutic applications.

Their ability to penetrate the cell nuclei and interact with critical cellular components like nucleosomes, histone proteins, and our DNA itself is groundbreaking! This interaction mechanism is a gateway to regulating our gene expression through direct engagement with the DNA structure - or indirectly via modifications in DNA methylation. Epitalon, as a tetrapeptide and short peptide, has been shown to have a direct effect on the activation of fibroblasts. 58] Fibroblasts are a type of cell found in connective tissue throughout the body, playing a crucial role in maintaining skin health and are primarily responsible for producing collagen and elastin.

Epitalon has recently been shown to have very powerful effects as a protector against oxidative stress and in its potential to combat accelerated aging due to ultraviolet (UV) radiation exposure. A study in 2022 examined Epitalon's influence on human skin fibroblasts and found that it appears to strengthen their defense mechanisms against environmental assaults - notably UV radiation! This study revealed that Epitalon stimulates the production of crucial antioxidant genes within the fibroblasts of our skin, namely NQO1, SOD1 (Superoxide Dismutase 1), and CATALASE.

An increase in our gene expression by a magnitude of 2.7, 2.6, and 3.2 times, respectively, enhanced the ability of fibroblasts to ward off oxidative damage. This gene expression was thought to occur through the Keap1/Nrf2 signaling pathway, which is a critical system for cellular defense against oxidative pressure.

UV radiation is recognized as a principal environmental cause of oxidative stress and accelerated aging in our skin. It was observed that, while the skin fibroblasts' inherent responses to UV exposure include the elevation of SOD1 and TXNRD1 genes, these reactions lacked alterations in NQO1 and CATALASE expressions. Here, Epitalon's ability to significantly upregulate these genes indicates a strong potential to enhance the skin's resistance against the harmful effects of UV – more than what these cells are naturally capable of. Aside from the immediate impact on our gene expression, the broader geroprotective and antioxidant roles of Epitalon may be a foundational key in combating the accelerated aging processes in dermal fibroblasts! "Geroprotective" here refers to the ability to prevent, slow down, or even reverse the process of aging, thereby extending the healthy lifespan of an organism.

By amplifying the antioxidant defense system within skin fibroblast cells, Epitalon could be a powerful, therapeutic approach to reverse the damaging effects of environmental stressors like UV radiation and might help delay the molecular, cellular, and visible markers of aging skin. 59]

Beyond their role in providing structural support, fibroblasts also assist in the skin's healing process by repairing wounds and generating new tissue. Given their function, maintaining healthy fibroblast function is essential for sustaining the skin's overall health, resilience, and youthful appearance.

Epitalon and the Repair
and Regeneration of Skin Cells

Recent research has revealed an intriguing way in which Epitalon engages with our DNA. Epitalon has been shown to have an effect on histones, particularly the variants H1/6 and H1/3. Histones are critical for our DNA packaging and regulation of our gene expression and act as spools that wrap around DNA. A study of 2020 found that Epitalon's ability to bind with these specific histones could significantly alter the genetic landscape within cells, potentially steering the gene expression patterns. This mechanism does not alter the DNA sequence but rather influences the cellular machinery's ability to access these genes. Consequently, this can lead to changes in how specific proteins are produced.

The study revealed that Epitalon boosts gene expression and the protein synthesis involved in neurogenic differentiation within human gingival mesenchymal stem cells (hGMSCs) as mentioned earlier. While the primary focus of this study was on neurogenesis, the broader potential of Epitalon to stimulate tissue repair and regeneration is exciting! 60]

Scientific Studies
on the Efficacy of Epitalon for Better Skin Health

Other studies also suggest that Epitalon may have potential benefits in reducing wrinkles, fine lines, and other signs of aging. One study investigated the impact of LK peptides and Epitalon, in concentrations ranging from 0.05-2.00 ng/ml, on the proliferation of organotypic skin cell cultures in young and old animals. LK peptides, or "leukokinin peptides," are a class of bioactive peptides that play a significant role in the regulation of blood pressure, inflammation, and pain responses in the body. They are part of the kinin peptide family and are involved in various physiological processes, including vasodilation, blood vessel permeability, and smooth muscle contraction.

These peptides and Epitalon were found to stimulate skin fibroblast proliferation by 29-45% in the skin cell cultures of both young and old rats. 61]

In another interesting study, it was discovered that male rats exposed to constant or natural light in North-Western Russia experienced an accelerated death rate, compared to those exposed to a standard 12-hour light cycle. Further, constant illumination was associated with an increase in the development of spontaneous tumors. Interestingly, the administration of Epitalon did not significantly alter the average lifespan of these rats. However, it did result in a significant normalization of the aging rate and the time of mortality rate doubling in groups exposed to either natural or constant illumination. Most interesting in terms of skin health, this study found that Epitalon significantly reduced the development of spontaneous tumors! Particularly testicular leydigomas and leukemia in the rats exposed to any light regimen.

In terms of how this relates to skin health, these findings may suggest that environmental factors, like light exposure, can have a very negative effect on our overall health, which in turn directly impacts the health of our skin. Therefore, we can deduce that peptides like Epitalon may potentially help regulate the accelerated aging and decline in health resulting from the constant exposure to artificial light. Either way, it is certainly a point of interest! 62]

Additionally, yet another study showed that Epitalon stimulates endogenous melatonin synthesis as the body ages, which could further help reduce the production of reactive oxygen species (ROS) in skin fibroblasts. 63] This supports the notion that Epitalon could be a powerful and promising agent in maintaining our skin's health, keeping it youthful and vibrant as we age!

Epitalon has also demonstrated very promising anti-inflammatory properties in various studies by modulating the immune response and targeting the production of pro-inflammatory cytokines.

Cytokines are small proteins that are released by a variety of cells within the body, primarily by immune cells such as T cells, B cells, macrophages, and mast cells.

Additionally, non-immune system cells like fibroblasts and endothelial cells are also able to secrete cytokines in response to various stimuli. These small proteins are instrumental in cell signaling, playing key roles in the body's inflammation response and in activating and directing immune cells to areas of infection, injury, or inflammation. Cytokines have a certain impact on the interactions and communications between our other cells. Pro-inflammatory cytokines, especially, are primarily responsible for promoting inflammatory responses. Epitalon's capability to reduce their production could make it an effective resource to manage the inflammatory processes overall. 63]

It has been shown to provide a soothing effect on the skin by reducing redness and swelling. This could in turn possibly lead to a decrease in the severity of certain conditions like acne, eczema, and psoriasis and provide relief to those who suffer from these skin conditions. Furthermore, Epitalon has also been shown to have a positive effect on the secondary damage which inflammation can cause. Chronic inflammation is well-known for causing tissue damage and worsening skin degradation, which may further contribute to signs of premature aging.

Better Skin – Better Health

In our endless pursuit of youthfulness and vitality, we have long been drawn to the promise of cosmetic enhancements, chemical concoctions, and an array of supplements - all promising to maintain our skin's youthful vibrancy and appearance. Far too often, though, we end up disappointed or left wanting for the natural glow and aliveness of the skin we had in our youth.

Epitalon may offer us a powerful regenerative modality to defy the outward signs of aging and achieve a profound level of cellular repair and rejuvenation!

What makes Epitalon so extraordinary is its ability to replicate the natural biological processes that diminish as we age. Instead of simply masking the signs of aging with creams or ointments, this peptide can activate the skin's own repair and renewal processes.

Beyond its cosmetic benefits, the influence it has on our skin's health has far-reaching implications in enhancing the skin's ability to function correctly as an organ. When we have healthier, more vibrant skin, the body is better equipped to handle immunity and detoxification. Meaning, we would not just "look better" but experience an overall increase in our health. Ultimately, in the field of skincare and anti-aging, this symbolizes a move towards methods that are more regenerative and holistic, rooted in an intricate understanding of our biology.

CHAPTER 11

VISION AND OCULAR PROTECTION

Our vision is one of the most critical senses we have, enabling us to fully experience the world around us. The sense of vision plays an important role in how we navigate, learn, and communicate with each other. Sight is a complex mechanism that involves not just the ability to see, but also the ability to understand and interpret what is seen. Given its importance, maintaining optimal eye health is essential for preserving our quality of life and independence as we age. Protecting the eyes, therefore, becomes a fundamental aspect of our overall health and well-being.

According to the World Health Organization, vision impairment is a significant global health concern that affects at least 2.2 billion individuals worldwide, with a substantial portion of these cases being preventable. Alarmingly, over 1 billion people suffer from vision impairment that could have been either prevented or which has not yet been treated. 65]

The primary causes of vision impairment and blindness on a global scale are refractive errors and cataracts. The economic implications of vision loss are staggering, with the annual global cost of lost productivity estimated at $411 billion. This figure not only reflects the individual challenges faced by those struggling with vision impairment but also emphasizes the broader socioeconomic impact affecting our society at large. 65] Age plays a significant role in the prevalence of vision impairment, with the majority of those affected being over the age of 50.

Common Age-related Vision Decline and Diseases

As we age, our bodies undergo various changes, and our eyes are no exception. Among the most common age-related vision disorders are presbyopia, macular degeneration, and glaucoma. These conditions not only reduce the quality of life for many elderly individuals, but also pose significant challenges for our healthcare system.

Presbyopia is a condition that typically begins to affect a person when they reach their 40s or 50s. It is characterized by a gradual loss of the eye's ability to focus on objects nearby, making actions like reading increasingly difficult. The primary cause of presbyopia lies in the hardening of the lens inside the eye. As we age, the crystalline lens becomes less flexible, which reduces its ability to change shape and focus on objects near to us through a process known as accommodation.

Age-related Macular Degeneration (AMD) is another major concern, affecting the central part of the retina known as the macula. The macula is responsible for our sharp and detailed vision, for instance when we drive. AMD can be classified into two types: "dry" and "wet" AMD. The dry form, which is more common, involves a gradual thinning and deterioration of the macula. This is often attributed to the aging and thinning of macular tissue or the deposition of pigment within the macula.

The less common wet form is more severe and involves the growth of abnormal blood vessels under the retina. This, in turn, leads to leakage and scarring that can result in rapid vision loss. The underlying causes of AMD relate to a combination of genetic, environmental, and lifestyle factors, including smoking, diet, and prolonged exposure to sunlight.

Glaucoma is a group of eye conditions that damage the optic nerve. This nerve is crucial for our vision, and this type of damage is often caused by abnormally high pressure inside the eye, known as intraocular pressure. This increase in intraocular pressure can result from a buildup of aqueous humor, which is a form of fluid that flows throughout the inside of the eye.

Normally, this fluid exits the eye through a drainage system. However, in people with glaucoma, this system becomes blocked. Fluid then accumulates which causes the internal pressure to build.

Causes and Development of Vision Decline

This deterioration in eye health and function can be attributed to various factors, such as oxidative stress, inflammation, genetic predispositions, and environmental influences.

Oxidative Stress

One of the primary mechanisms leading to age-related vision decline is oxidative stress. The eye, particularly the retina, is highly susceptible to oxidative damage due to its high metabolic rate, prolonged exposure to light, and rich content of polyunsaturated fatty acids. Oxidative stress occurs when there's an imbalance between the production of reactive oxygen species (ROS) and the body's ability to neutralize these harmful compounds through antioxidants.

Over time, the accumulation of oxidative damage can impair cellular function and lead to the death of retinal cells. This mechanism is notably involved in the pathogenesis of age-related macular degeneration (AMD), where oxidative stress contributes to the degradation of the macula, resulting in significant vision loss.

Inflammation

Chronic inflammation is another key factor in the development of age-related eye diseases. With aging, the immune system's regulation becomes less efficient, leading to an increase in systemic inflammation. In the eye, this can manifest as an overactive immune response, causing damage to delicate ocular tissues. With AMD, for instance, inflammation leads to the accumulation of drusen (lipid-rich deposits) beneath the retina, contributing to retinal damage and vision loss. Similarly, in glaucoma, inflammatory processes can exacerbate optic nerve damage, further impairing vision.

Genetic Predispositions

Genetic factors significantly influence the risk of developing age-related eye conditions. Numerous studies have identified specific gene variants associated with increased susceptibility to diseases like AMD, glaucoma, and diabetic retinopathy. For example, variations in the complement factor H (CFH) gene and other genes involved in the immune response have been linked to a higher risk of AMD. These genetic predispositions can affect various biological pathways, including those related to inflammation, oxidative stress, and cellular metabolism, thereby modulating an individual's vulnerability to vision decline.

Environmental and Lifestyle Factors

Environmental and lifestyle factors also play a crucial role in the health of aging eyes. Factors such as smoking, poor diet, and excessive exposure to ultraviolet (UV) light can exacerbate oxidative stress and inflammation, thereby accelerating the progression of eye diseases.

Smoking has been identified as a significant risk factor for AMD and cataracts, likely due to the direct toxic effects of tobacco smoke on ocular tissues. A diet lacking essential nutrients and antioxidants can weaken the body's natural defenses against oxidative damage, while prolonged UV exposure can cause cumulative harm to the retina and lens.

Integration with Regenerative Aging

The approach of regenerative therapies for anti-aging holds significant promise for addressing age-related vision decline. The connection between these disorders and degenerative aging lies in the underlying mechanisms of cellular and tissue deterioration that occur as part of the aging process. Regenerative therapies seek to address these root causes by promoting the repair and regeneration of damaged tissues, thereby restoring vision or preventing further decline.

Stem Cell Therapy:

One of the most promising interventions in these therapies is the use of stem cell therapies. Stem cells can uniquely differentiate into various types of cells, offering the potential to replace or repair damaged retinal cells. For instance, research is now focused on using pluripotent stem cells to generate retinal pigment epithelium (RPE) cells, which play a critical role in supporting retinal function. Transplanting these cells into the retina could help restore vision in patients with AMD and other retinal diseases.

In my book, "How To Reverse Aging- A Comprehensive Guide To Copper Peptides," it is discussed how the GHK-Cu peptide activates gene expression so that pluripotent stem cell production increases.

Gene Therapy:

Gene therapy represents another cutting-edge approach to anti-aging in the context of vision decline.

By targeting the genetic factors of eye diseases, gene therapy aims to correct or mitigate the effects of mutations contributing to conditions like retinitis pigmentosa, and Leber congenital amaurosis. Techniques such as CRISPR-Cas9 gene editing are now examined for its ability to precisely edit the DNA of ocular cells. This could potentially restore its normal function and prevent further deterioration.

Neurotrophic Factors and Biologics:

The use of neurotrophic factors and biological agents is also being investigated as a means to promote the survival and health of retinal cells. These substances support the natural repair mechanisms within the eye, reducing cell death and fostering regeneration. For example, the injection of growth factors or the use of encapsulated cell technology to deliver these factors continuously could offer new ways to treat or slow the progression of glaucoma and AMD.

Epitalon's Protective Mechanisms for the Eyes

While Epithalamin is extracted directly from the pineal gland, Epitalon is created in a laboratory setting to mimic the effects of its natural counterpart. This synthetic version offers a more accessible and controlled way to study and utilize the potential benefits related to anti-aging, cellular health, and the regulation of various bodily functions.

Its potential protective mechanisms for the eyes is an area of heightened interest, especially as many populations worldwide face increasing age-related ocular conditions.

Antioxidative Properties

As mentioned earlier, the eyes are particularly vulnerable to oxidative stress. Oxidative stress occurs when there is an imbalance between free radicals and antioxidants in the body. Free radicals are oxygen-containing molecules with an uneven number of electrons, which allows them to easily react with other molecules. That is the reason why they are also called "unstable molecules". While the body naturally produces free radicals as a byproduct of cellular processes, excessive amounts of them causes cellular harm through the process of oxidative stress. This type of stress damages cells, proteins, and our DNA, leading to various diseases and classical signs of degenerative aging.

Epitalon exerts its protective effects on eye health through its antioxidative properties. It helps neutralize free radicals, thereby reducing oxidative stress, and its damaging effects on retinal cells. Studies have demonstrated that both Epithalamin and Epitalon not only exceed some of the antioxidant effects of melatonin, but also offer a unique mechanism of action in combating oxidative stress. 66]

Epitalon and Epithalamin have been shown to possess remarkable antioxidant capabilities by enhancing the body's natural defense system against reactive oxygen species (ROS). This is achieved through the direct scavenging of ROS and the stimulation of the body's antioxidant enzymes, like superoxide dismutase (SOD), glutathione peroxidase, and glutathione-S-transferase. These enzymes play a crucial role in neutralizing free radicals and protecting cells from oxidative damage, which is a key factor in the aging process.

Moreover, this research highlighted the fact that Epitalon does not merely mimic the antioxidant effects of melatonin but also stimulates the production of melatonin, which itself is a potent antioxidant produced by the pineal gland.

This dual action contributes to Epitalon's ability to fortify our antioxidant defense system more effectively than melatonin alone. One of the unique aspects of Epitalon's antioxidant mechanism is its potential to stimulate the expression of antioxidant enzymes that interact with various ROS through the binding of transition metals like iron (Fe^{2+}). This suggests that Epitalon can interrupt oxidation chains, further enhancing its antioxidant capacity.

The collective results from these studies suggest that peptides from the pineal gland, particularly Epithalamin (Epitalon), significantly bolster the body's antioxidant defenses. This, in turn, contributes to their geroprotective (aging-delaying) properties, making them a promising avenue for research and application in anti-aging therapies and overall health maintenance.

Effects on Telomere Activity

In the context of the restoration of telomere length discussed earlier, Epitalon's effect on telomere length, as it applies to eye health, may help maintain or even extend telomere length in ocular cells, and delay the onset of age-related eye diseases. 67]

Physiological Benefits of Epitalon for Eye Health

Epitalon has also shown great promise in the field of eye health due to its ability to affect cell function and growth. This peptide has been found to play a significant role in the health of the pineal gland and the retina, which is the part of the eye that is responsible for our vision. Stimulation of Gene Expression and Protein Synthesis during Neurogenesis.

What makes Epitalon a promising influence on ocular health is its capacity to facilitate the transformation of certain cell types into neuronal cells.

Neuronal cells are crucial for the proper functioning of the retina, which is often compromised in eye diseases such as macular degeneration or glaucoma. By promoting the regeneration of damaged cells within the retina, Epitalon offers a groundbreaking approach to treating these conditions.

Its potential to encourage damaged retinal cells, to regenerate or even transform them into functional neuronal cells, holds promise for restoring vision or preventing further degeneration in those affected by these eye diseases.

In a study entitled, "AEDG Peptide (Epitalon) Stimulates Gene Expression and Protein Synthesis during Neurogenesis: Possible Epigenetic Mechanism," it was found that Epitalon can increase the expression of neurogenic differentiation markers such as Nestin, GAP43, β Tubulin III, and Doublecortin. These markers have a critical role in neural growth, and the establishment of neural networks, which are essential for the proper functioning of the retina.

Furthermore, the peptide's interaction with histones H1/3 and H1/6 suggests an epigenetic mechanism of action, where Epitalon may modify the expression of genes involved in neuronal health and development. Histones are essential proteins that significantly contribute to DNA organization and functionality within cells.

Structurally, histones work to compact the DNA by coiling it around themselves, facilitating its accommodation within the limited space of the cell nucleus. This compact form is critical not only for spatial efficiency but also for the regulation of gene expression. By selectively exposing or shielding specific DNA regions, histones play a pivotal role in controlling which genes are active at any given time, thereby managing cellular functions. This epigenetic regulation could lead to enhanced repair capabilities of damaged retinal cells and support the maintenance of healthy retinal function. 68]

Other Clinical Evidence
Supporting Epitalon's Efficacy

Research has been done on the effects of Epitalon on patients with unstable glaucoma, showing very promising results! In a study, it was administered to 28 patients with the aim of normalizing the intraocular pressure (IOP), which is a critical factor in managing glaucoma. Epithalamin has been shown to exhibit a regulatory effect on the functional activity of the sympathetic-adrenal system, and the adrenal cortex. The clinical efficacy of the administration of Epitalon proved to be beneficial in its ability to normalize the biochemical activity within the neuro-humoral triad, which includes adrenaline, norepinephrine, and hydrocortisone.

This normalization led to significant physiological changes that were beneficial for eye health, such as an increased diameter of arterioles in the bulbar conjunctiva, and the optic disk. Furthermore, a notable reduction in the degree of congestive angiopathy and intravascular aggregation of erythrocytes was determined. 69] These findings suggest that Epitalon may not only help manage IOP in patients with unstable glaucoma, but that it may also improve the overall vascular health of the eye.

Epitalon's Impact on Other Visual Conditions

Epitalon has been shown to have regenerative properties that could be very beneficial in a range of ocular conditions, particularly those that stem from or result in the degeneration of retinal cells.

Retinitis Pigmentosa

Retinitis pigmentosa (RP) refers to a collection of inherited eye disorders that primarily affect the retina, which is the light-sensitive layer at the back of the eye responsible for converting light into neural signals for the brain, to interpret visual images.

This group of conditions leads to alterations in how the retina responds to light, progressively impairing vision.

The hallmark symptom of RP is the difficulty patients have with impaired night vision or seeing in conditions of low light. As the condition progresses, a person experiences a gradual narrowing of their field of vision (also known as "tunnel vision"), alongside problems with color vision and central vision in later stages. RP is caused by genetic mutations that are passed down through families, affecting the photoreceptor cells in the retina. These cells, known as "rods and cones," are critical in night, peripheral, and color vision.

While there is currently no known cure for RP, conventional strategies aim to maximize any remaining vision and improve a person's quality of life. In the study by Vladimir Khavinson et al. (2002), they discovered that the mechanism by which Epitalon improves the condition of the retina in retinitis pigmentosa involves its ability to bind to pineal regulatory element (PIRE) sequences in our DNA. These sequences are unique promoters of genes expressed in the retina, including genes for photoreceptors, and the gene for arylamine N-acetyltransferase, which is an enzyme critical for melatonin biosynthesis.

By binding to these specific DNA sites, or to transcription factors regulating gene expression, Epitalon positively affects melatonin production in the pineal gland and retinal functions. This interaction suggests that Epitalon can help preserve the structure and function of the retina, offering a promising therapeutic strategy for retinal degenerative conditions. 70]

Diabetic Retinopathy

Diabetic retinopathy is a condition that affects people with diabetes. This condition damages the blood vessels of the retina, leading to vision impairment and, in severe cases, blindness.

The primary cause of diabetic retinopathy is prolonged high blood sugar levels. This leads to leakage, swelling, and sometimes the growth of abnormal new blood vessels on the surface of the retina.

Symptoms include seeing spots or dark strings floating around in your vision (floaters), blurred vision, fluctuating vision, dark or empty areas in your vision, and actual vision loss. Diabetic retinopathy progresses in stages, from mild nonproliferative retinopathy to more advanced proliferative retinopathy, where the risk of vision loss increases significantly.

In a research project done by Trofimova and Khavinson (2001), 104 individuals suffering from diabetic retinopathy received treatment involving Retinalamin, Epithalamin, and Cortexin (derived from the brain's cortex). For the majority of these patients, this combination of bio-regulatory peptides not only enhanced their vision but also effectively halted the bleeding and fluid leakage, minimizing macular swelling. Furthermore, this treatment was found to consistently improve the retina's functionality and circulation, with no negative side effects reported. 71]

Considerations for Epitalon Usage in Eye Health

When considering Epitalon for eye health, several factors should be considered, such as dosage, potential interactions with other medications, and overall safety.

Dosage Considerations

The appropriate dosage of Epitalon can vary widely depending on individual factors such as age and weight. While research on the optimal dosing for ocular health is still emerging, it is crucial to begin with the lowest possible dose that might be effective, and to monitor for any adverse reactions.

Potential Interactions

Before administering Epitalon, potential interactions with other medications or supplements should be considered. Epitalon's mechanism of action may influence various cellular processes, which could interact with the pharmacological actions of other compounds. For instance, if someone is on medication that affects cell replication or immune function, Epitalon could theoretically alter the effectiveness of these treatments or lead to unexpected outcomes.

Safety and Side Effects

Although more research on long-term administration of Epitalon has yet to be studied, overall, preliminary research has shown it to be well-tolerated.

The Importance of Professional Consultation

Consulting with a professional before beginning any treatment is essential. It is not recommended to self-dose or self-treat without appropriate guidance. A professional can recommend appropriate dosages and monitoring to mitigate any possible side effects.

Regulatory Considerations

Also, the availability and regulatory status of Epitalon varies by country. Some areas only offer Epitalon for laboratory use, or for research purposes. Ensure that the Epitalon you purchase is obtained from a reputable source of good quality.

CHAPTER 12

BONE AND JOINT HEALTH

As we age, significant changes occur in our bones and joints that impact our posture and gait. With aging, many experience a loss of bone mass or density, particularly post-menopausal women due to a decrease in calcium and other essential minerals. The spine also undergoes notable changes; the vertebrae lose much of their mineral content, and the discs between them become thin. This can result in a shortened upper body (trunk), a curved spine, and potentially the growth of degenerative bone spurs.

Additionally, our joints become stiff and less flexible. The fluid which normally enables smooth movement may decrease, and the cartilage preventing the bones from rubbing together begins to wear away. This wear and tear can lead to osteoarthritis. Additionally, minerals may also be deposited around the joints and cause further stiffness and discomfort. These traditionally recognized age-related deteriorations of the bones and joints significantly affect a person's posture and walking pattern, leading to slowed movements and increased weakness.

Between 2019 and 2021, a survey called the National Health Interview Survey (NHIS) found that about 53.2 million adults in the US — or 21.2% — were diagnosed with arthritis or a similar condition, such as rheumatoid arthritis, gout, lupus, or fibromyalgia. 72] This shows that many individuals in the US deal with these health issues. The survey also found that more women (24.2%) seem to develop these conditions compared to men (17.9%).

Issues such as these notoriously increase with what we consider traditional aging. Out of those aged 45 to 64, 26.0% were found to suffer from similar conditions. Out of those 65 years and older, almost half of these cases (47.3%) confirmed being diagnosed with arthritis. 72]

These numbers highlight the need for effective preventative measures and the need to find the best ways to holistically manage and reduce their impact on our health.

Common Bone and Joint Issues Due to Aging

As we age, the conditions of joint and bone disease not only compromise our skeletal health but also impede our mobility and impact the quality of our life.

Osteoporosis

Osteoporosis is a condition characterized by a decrease in bone density and mass, making the bones fragile and more prone to fractures. This condition primarily results from an imbalance in the bone remodeling process, a lifelong process where mature bone tissue is removed (resorption), and new bone tissue is formed (ossification).

In young individuals, these processes are well-balanced. However, with degenerative aging, ossification becomes less efficient compared to resorption, leading to a net loss in bone mass. Factors contributing to this imbalance include hormonal changes, particularly a decrease in estrogen in post-menopausal women, reduced physical activity, and inadequate intake of calcium and vitamin D. The spine, hip, and wrist are the most commonly affected areas, where fractures significantly impact an individual's mobility and independence.

Osteoarthritis

Osteoarthritis (OA) is the most prevalent form of arthritis, characterized by the degradation of joint cartilage and the underlying bone. The primary mechanism behind OA is the breakdown of the cartilage. This is the tough, elastic tissue that covers and protects the ends of the bones within a joint. Cartilage degradation leads to a reduction in joint space and increased friction between the bones, resulting in pain, stiffness, and decreased mobility. Although OA can be attributed to a variety of factors, such as genetics, obesity, and wear and tear, aging itself decreases the body's ability to repair cartilage efficiently, which worsens the progression of OA in older adults.

Degenerative Disc Disease

Degenerative Disc Disease (DDD) refers to changes in the spinal discs associated with degenerative aging. Spinal discs are soft, compressible discs, separating the interlocking vertebrae that make up the spine. These discs serve as shock absorbers for the spine, allowing it to flex, bend, and twist. They also provide space between the vertebrae so that our spinal nerves can pass out of the spinal canal and to the proper places in the body and organs. With age, these discs can lose their hydration and elasticity, which decreases their natural gel-like substance. In essence, they dry out.

Additionally, the outer layer of the disc can develop "cracks," leading to disc herniations or bulging. This process causes pain, numbness, or weakness in the neck or back. Factors like our genetics and lifestyle habits such as smoking influence the severity and progression of DDD. Smoking increases the incidence of DDD in the cervical spine by a factor of 10!

Symptoms of Age-Related Bone and Joint Issues

Age-related bone and joint issues manifest through a variety of symptoms that significantly impact a person's quality of life. These symptoms often develop gradually and worsen over time as the conditions progress.

Pain:

Pain is a common symptom associated with various age-related bone and joint conditions. In osteoarthritis, the pain results from the cartilage being broken down within the joint, leading to increased friction and pressure. This causes an achy or sharp pain when the joint is moved or at rest. In the case of osteoporosis, this type of pain often comes from fractures that occur in weakened bones, particularly in the spine, hip, or wrist. Even minor falls or stresses can lead to fractures, causing acute or chronic pain.

Stiffness:

Joint stiffness is another hallmark symptom, especially with conditions like osteoarthritis. It typically occurs after periods of inactivity or rest and may lessen slightly with movement. The stiffness results from changes in the synovial fluid which lubricates the joints, making movements less smooth and more restricted.

Reduced Range of Motion:

Degenerative aging leads to a reduced range of motion of the joints, primarily due to the degradation of joint structures and the development of bone spurs, such as with osteoarthritis. These changes limit flexibility and mobility, making it difficult to perform activities requiring us to bend, twist, or stretch. Over time, a person might notice that they are less able to reach over their head, bend down to reach for an item on the floor, or even walk as easily as they once could.

Common Treatments for Bone and Joint Issues

Many of these conditions are considered inevitable as we age, with traditional treatments aiming to manage symptoms, reduce pain, and improve mobility as best they can.

Medication

For osteoporosis, bisphosphonates are commonly prescribed to try and prevent bone loss and reduce fracture risk by inhibiting osteoclasts which are cells that break down bone tissue. Denosumab, another drug, works by targeting a different pathway to slow down bone resorption.

Hormone-related therapies, such as estrogen and selective estrogen receptor modulators (SERMs), may also be employed. For osteoarthritis, over-the-counter pain relievers like acetaminophen and NSAIDs are often used with the aim of alleviating pain and inflammation. In more severe cases, corticosteroid injections directly into the joint may be used to provide temporary relief.

However, these medications are not an actual handling of the condition and all come with a range of side effects. NSAIDs, for example, may cause gastrointestinal issues such as ulcers or bleeding, especially with long-term use. Bisphosphonates have been associated with rare but serious complications, such as osteonecrosis of the jaw or atypical femoral fractures. Some hormone-related therapies carry an increased risk of blood clots or stroke.

Physical Therapy

Physical therapy is one of the most used therapies to try and manage age-related musculoskeletal issues. Tailored exercise programs strengthen muscles and may improve flexibility or enhance joint mobility. Ironically, movement can reduce the pain related to some forms of arthritis, for instance, but again offer no permanent solution.

Weight-bearing and resistance exercises can be particularly beneficial for those struggling with osteoporosis, as it helps build bone density.

Surgical Interventions

In cases where medication and physical therapy "no longer work", surgical interventions may be recommended. An example of this may be osteoarthritis progressing to the point of a "bone on bone" condition. Here, there is no cartilage left to protect the bone ends, making the joint unstable and causing pain.

Joint replacement surgery, like hip or knee arthroplasty, is a common suggestion for advanced osteoarthritis. This procedure involves replacing the damaged joint surfaces with artificial components.

It is important to understand how important these interventions are when pain in the knee or hip is affecting a person's ability to squat their own body weight! The number one "last straw" tripwire that sends a person into nursing care is the inability to do this because, by the time this happens, in many cases, a person's spouse/partner is also old and cannot assist in helping their loved one to get on and off of the toilet and it can actually become dangerous for both!

Additionally, spinal surgeries such as vertebroplasty or kyphoplasty may be offered to stabilize spinal fractures. Both procedures aim to stabilize the fractured vertebra, alleviate pain, and restore mobility, but come with their own risks.

Vertebroplasty involves the percutaneous injection of a medical-grade bone cement into the fractured vertebra. A needle is guided into the damaged vertebra using imaging techniques, like fluoroscopy, under local or general anesthesia. Once positioned, bone cement is injected to stabilize the fracture. Kyphoplasty also involves the injection of bone cement but includes an additional step: before cement injection, a small, specialized balloon is inserted and inflated within the fractured vertebra, creating a space.

This process aims not only to stabilize the fracture but also to restore some of the lost vertebral height, and reduce spinal deformity. After the balloon is removed, the cavity is filled with bone cement.

Post-surgery issues, complications, and problems for both procedures can include:

- Infection:
  As with any surgery, there is a risk of infection at the injection site or within the treated vertebra.

- Increase in pain:
  Some patients experience an initial increase in back pain following the procedure.

- Cement leakage:
  One of the most common complications is the leakage of bone cement into the surrounding areas. This can lead to nerve damage, spinal cord compression, or the need for further surgeries.

- Allergic reactions:
  Some may experience allergic reactions to the materials used, like the bone cement or contrast dye.

- New fractures:
  Some evidence suggests that the increase in vertebral stiffness following the procedure can lead to adjacent vertebrae fractures.

The Role of Epitalon
in Maintaining Skeletal Health

With advances in technology and a deeper knowledge of the human body, down to the molecular level, we have seen some amazing new treatments develop in the field of medicine. In the context of bone health, Epitalon stands as one of the best peptides to promote longevity, for its potential to enhance bone density and overall vitality. 73]

One notable study by Khavinson et al. (2020) investigated the mechanism by which Epitalon stimulates gene expression related to osteoblastic activity and proliferation. This research provided us with valuable insights into how peptides like Epitalon can regulate the functional activities of cells involved in bone health, including their proliferation, differentiation, and apoptosis. 74]

Effects of Epitalon on Bone Density

Joint health is closely related to our bone health, as healthy joints enable smooth and efficient movement. This is essential to keeping our bones aligned, and minimizing undue stress which can lead to wear and tear. Moreover, the integrity of our joints directly impacts our ability to engage in physical activity, which is a key factor in promoting bone density, and preventing conditions like osteoporosis. As joints deteriorate, the reduced mobility and increased discomfort not only compromise bone health by limiting physical activity but alters the mechanical load which affects their strength and structure.

One of the key mechanisms through which Epitalon may exert its effect is through the modulation of the way the pineal gland functions. 75] The pineal gland is crucial in regulating our circadian rhythm and, indirectly, various hormonal activities that influence our bone metabolism.

For instance, melatonin, which is produced by the pineal gland, has been associated with bone health, suggesting that Epitalon's impact on the pineal gland and melatonin production could affect the regulation of bone density.

Moreover, the general antioxidative and tissue-regenerative properties of Epitalon may play a role in supporting bone health. 75] Oxidative stress and inflammation are contributing factors to the loss of bone density so, by potentially reducing oxidative damage and promoting cellular repair, Epitalon may indirectly support the maintenance or improvement of bone mineral density.

Effects of Epitalon on Joint Health

Our joint health is central to the well-being of the entire musculoskeletal system. Moreover, the integrity of our joints directly impacts our ability to engage in any physical activity. As joints deteriorate, whether due to age, injury, or conditions like osteoarthritis, the reduced mobility and increased discomfort not only compromise bone health, but also increase the risk of bone-related diseases.

One of the key mechanisms through which Epitalon may offer benefits for joint health is through its ability to increase the synthesis of collagen and glycosaminoglycans. 76] Collagen is not only important for our skin's health but is also a fundamental component of cartilage.

Glycosaminoglycans are large polysaccharides that are critical in maintaining the structural integrity of cartilage. They provide resistance to all forms of compressive forces, so by stimulating the production of these molecules, Epitalon can help maintain the resilience and function of joint cartilage. This can potentially slow down — or reverse — any degenerative joint alterations associated with traditional aging.

Moreover, the stimulation of decorin synthesis by Epitalon is another way by which it may support our joint health. Decorin is a small leucine-rich proteoglycan that interacts with collagen fibrils, regulating the quality and organization of collagen fibers. This interaction is vital in maintaining the structural integrity and mechanical properties of cartilage. This suggests that an increase in decorin synthesis could contribute to healthier, more resilient joints!

The promotion of angiogenesis (the formation of new blood vessels), and nerve outgrowth by Epitalon also supports its broader regenerative potential. The promotion of improved blood flow to joints enhances their nutrient delivery, waste removal, and supports repair processes. Similarly, the promotion of nerve outgrowth may help maintain the functionality and health of joints. By mitigating oxidative stress and inflammation, Epitalon might also help protect joints from damage and degeneration caused by wear and tear. 75]

Reduced Risks of Osteoporosis with Epitalon

The positive effect of Epitalon on telomerase and the length of our telomeres also suggests that Epitalon may mitigate age-related degenerative processes, including those affecting our bone health. The significance of bone density in osteoporosis cannot be overstated, as it directly correlates with the strength and structural integrity of our bones. Research suggests that Epitalon could promote bone health by potentially influencing the biological mechanisms that aging and degeneration are based on. 77]

Studies Supporting the Benefits of Epitalon for Bone Health

In the study done by Vladimir Khavinson and his team, they examined how Epitalon affects human gingival mesenchymal stem cells (hGMSCs). 78] They were particularly interested in seeing if Epitalon could help these cells turn into nerve cells, which is important for fixing damaged nerves, and treating diseases that affect the brain and neurons. Their research showed that when hGMSCs were treated with Epitalon, it boosted the activity of certain genes, and the creation of the proteins needed to generate new nerve cells.

This suggests that Epitalon has a strong ability to have these stem cells turn into nerve cells. This is thought to happen through epigenetic changes, which means Epitalon might be able to control what genes are turned on or off, without changing the DNA itself.

Given its multifaceted benefits, Epitalon represents a promising future research in regenerative medicine. Studies have highlighted the potential of hGMSCs for bone regeneration, and research already examined the osteogenic capabilities of GMSCs in both lab settings and live models. Characterized by their quick growth and the presence of important stem cell markers, such as CD90, these cells stand out for their regenerative potential.

Laboratory studies demonstrated their ability to differentiate, while other studies showed their potential to repair bone defects. 76]

Overall, as research on Epitalon keeps advancing, it is becoming more and more obvious that Epitalon could become extremely important in anti-aging, and healing. By continuing to study what its functions are, and how these affect us, we may be able to fully harness the essence of its healing abilities. This could lead to new treatments that greatly improve life for those struggling with bone and joint issues, or entirely prevent them!

CHAPTER 13

DOSAGE AND ADMINISTRATION

Determining the correct dosage of Epitalon and the method of administration is essential for effective and safe application. Getting the dosage right ensures that we receive the precise amount needed to produce a therapeutic effect without leading to any negative effects or no results.

The administering of Epitalon can be done orally, through an IV, or subcutaneously. Additionally, the frequency and length of treatment also influence its effectiveness and safety. These crucial details are defined through rigorous clinical trials and research to enhance the therapeutic benefits. The bioavailability of peptides, their stability, and their capacity to trigger specific responses are influenced by the method of administration.

Administration of Epitalon:
Exploring Therapeutic Approaches

There are various methods of administering Epitalon, each offering its own unique benefits and considerations.

Subcutaneous Injections

Subcutaneous injections are widely used to deliver Epitalon directly into the body and finally into the bloodstream. This approach allows the peptide to be absorbed quickly and provides an immediate effect.

Here, the dosing is exact and is tailored to official recommendations. Many times, subcutaneous injections are preferred in a clinical setting.

Intramuscular Injections

Intramuscular injections, on the other hand, provide an alternative way of injecting Epitalon. By delivering Epitalon deep into muscle tissue, it allows for a slower release and absorption of the peptide into the bloodstream. Intramuscular injections can offer a balance between the rapid action of subcutaneous injections and the prolonged release that may be more suitable for sustained effects.

Oral Supplements

Oral supplementation offers a more non-invasive alternative to injections. While this might result in lower bioavailability due to the digestive process, it remains a popular choice for its ease of use and convenience. Oral supplements can be particularly appealing for the long-term administration of Epitalon, as they do not require the same level of supervision or repeated injections.

Topical Applications

Topical application targets the peptide's effects directly onto the skin. This type of administration leverages the peptide's properties by applying it directly to a desired area, which offers a focused approach to skin rejuvenation.

Intranasal Administration

Intranasal administration is yet another innovative method that has been explored. This method involves administering the peptide through the nasal passages, where it can bypass the blood-brain barrier and exert its effects directly on the central nervous system. This method is particularly intriguing for its potential to target neurological aspects of aging and neurodegenerative diseases, as it offers a very non-invasive way of influencing the brain.

<u>Phototherapy Patches</u>

Innovative, non-transdermal patches have also been developed, representing a breakthrough in biotechnology by activating the body's natural production of Epithalamin. By stimulating the body upon contact with the skin, these patches initiate a process that enhances the body's capability to generate these peptides. This method focuses on encouraging the body to elevate its peptide levels naturally. The benefits of this technology lie in its ability to target cellular renewal and telomere extension from within, which are fundamental to the promotion of longevity.

Products and Treatments Available with Epitalon

As research into Epitalon's effects continues to evolve, a variety of products and treatments have become available.

Injectable Formulations

One of the most common methods available is injectable formulations. These are offered in various concentrations, such as 10mg/mL, in 5ml or 10ml vials.

Peptide Blends:

Epitalon is also available in peptide blends, where it is combined with other peptides or agents, like NAD+ or mesenchymal stem cells. These are mixed to enhance their effects on overall health and are designed to target multiple pathways involved in the aging process. These mixes offer a synergistic effect that can amplify the benefits of Epitalon.

Research Chemicals:

These are intended for scientific or research purposes.

These products, such as Epitalon TFA, are intended for laboratory research to further understand the peptide's mechanisms and therapeutic applications. 79]

Telomerase Activators:

As a telomerase activator, Epitalon is marketed for its ability to extend the length of our telomeres. Some products feature Epitalon as a telomerase activator for its mechanism as an influence on telomerase.

Purity and Concentration:

Products of Epithalone (Epitalon) with a 99% purity grade in specific concentrations are also available. Such specifications are crucial for ensuring the reliability and reproducibility of research findings, enhancing the safety and efficacy of Epitalon.

Recommended Dosages of Epitalon for Different Age Groups and Conditions

Factors like age, existing health status, and specific wellness goals are all factors that determine the dosage used. On average, a common dosage of injected Epitalon ranges from 5 to 10mg per day. New users, however, are generally recommended to start at a lower dose and observe how their body reacts. Normally, Epitalon is administered in cycles, for instance, of 20-day intervals, followed by a break with a new cycle then starting. The correct dosage is something your healthcare provider would recommend and no recommendations are given here.

Young Adults (Ages 20-30)

Young adults who may use Epitalon primarily for its potential to enhance physical performance, immune function, or general well-being might have a recommended lower dosage at the start of their treatment. A cycle of 5mg per day for 20 days, followed by a break, could suffice. This age range typically has a robust natural production of the peptides Epitalon mimics, so a higher dose should not be needed. Consult with your provider for your exact dosages.

Middle-Aged Adults (Ages 31-50)

For individuals in this age range, dosages can range from 5mg to 7mg per day, administered in a 20-day cycle. The slightly increased dosage accounts for the natural decline in our production of peptides, beginning around this phase of life. Consult with your provider.

Older Adults (Ages 51 and Above)

For older adults, a dosage closer to the upper end of the spectrum, like 8-10mg per day, for a 20-day cycle, may be more beneficial. This takes into account the further decreased natural production of peptides due to aging and the increased need for support in various physiological functions. As mentioned above, consult with your provider.

Individuals with Specific Health Conditions

For those seeking Epitalon's therapeutic benefits for specific health conditions or goals, such as improved sleep or enhanced recovery, the dosages may significantly vary. In some cases, a customized cycle length and dosage might be necessary to align with specific health objectives and conditions.

Frequency of Administration
for Optimum Benefits

An optimal frequency of administration varies from person to person, influenced by our individual goals, physiological response, and existing health conditions.

Daily Administration:

For those seeking immediate effects, a daily administration of Epitalon can be considered, for a certain cycle. This regimen might be particularly effective for those focusing on improving the quality of their sleep or who are seeking a faster response. According to research published in the National Library of Medicine, Epitalon can play a role in protecting against post-ovulatory aging-related declines, suggesting its importance in time-sensitive health interventions. 80] Daily dosing over short periods, typically ranging from 10 to 20 days, can provide a concentrated exposure to the peptide, which may catalyze its beneficial effects.

Weekly or Monthly Cycles:

For general maintenance and gradual improvements in specific health markers, weekly or monthly cycles may be beneficial. This approach allows for a sustained presence of the peptide in the body, which supports ongoing physiological processes without overwhelming it. For instance, studies have explored the effect of Epitalon as a geroprotector (an element which affects the root cause of aging) in organisms like the common fruit fly, Drosophila melanogaster. This indicates that prolonged exposure to Epitalon may have a cumulative benefit. 81] An optimal frequency is always tailored to a person's individual needs and health objectives. Starting with a daily regimen until initial improvements are achieved, then transitioning to a less frequent dosing schedule for overall maintenance might be the most effective strategy.

Monitoring and Adjustments:

Regardless of the chosen frequency, close monitoring of the body's response and reactions is important. Adjustments to the dosing schedule might be necessary, depending on the initial reactions. Consulting with a professional who is experienced in peptide therapy provides valuable guidance in optimizing the frequency of Epitalon for the best therapeutic outcome.

Contraindications for Epitalon and Potential Risks

While Epitalon is celebrated for its minimal side effects and general tolerability, understanding and managing potential individual reactions is important for optimizing its use.

Variability in Individual Reactions

The body's response to any form of treatment can always vary from person to person. Minor injection site irritation is one of the more common reactions, as with any other injectable method. This irritation is typically mild and transient and resolves without the need for any intervention.

In addition to localized reactions at the injection site, some may experience systemic symptoms like headaches, fatigue, or a general sense of discomfort. These symptoms, while not that pleasant, are generally indicative of the body's initial adjustment phase to the introduction of the peptide. It is important to note that these reactions are reported by only a small handful of people and are usually short-lived.

Mitigation Strategies

Understanding that these reactions are typically temporary is key to managing them effectively. Here are several strategies that can help mitigate these initial reactions:

A Gradual Introduction

Starting with a lower dose of Epitalon, to gradually increasing it is a generally recommended approach that may help minimize any potential reactions. This allows the body to adjust more gently to the peptide.

Monitoring and Adjustment:

Keeping a log of any reactions and discussing these with a professional can be useful. This ongoing monitoring makes it possible for the provider to adjust the treatment plan, optimizing the therapeutic experience while minimizing any discomfort.

Injection Technique and Care:

Proper care of the injection site can significantly reduce the occurrence of irritation. Following best practices given to you by your provider on proper care is essential.

Proactive Risk Mitigation:

By tailoring the dosage to a person's specific needs and monitoring for any adverse reactions throughout the treatment, Epitalon can render many benefits! This personalized approach ensures the treatment aligns best with the wanted outcome.

Consultation and Care:

For those with pre-existing medical conditions or who are undergoing any other prescribed treatments, supervision by a professional is vital. This measure allows for a comprehensive assessment of any possible contradictions that may come up.

Safety Profile of Epitalon
and Lack of Side Effects

As mentioned, Epitalon is well-known for having very few - if any - side effects, and for being easily tolerated. The fact that Epitalon hardly has any negative reactions highlights its potential as a safer option for those who wish to reduce the signs of aging while keeping their comfort and safety in mind.

Easy to Tolerate with Few Side Effects

One of the key benefits of Epitalon is that it causes minimal reactions at the injection site when given under the skin. This is different from many other injectable treatments that may cause a lot of irritation or discomfort. Epitalon is renowned for its degree of tolerability. It is also well-known for not causing any gastrointestinal issues, which are sometimes a concern with oral treatments. This makes Epitalon a very attractive choice for long-term use.

Supported by Science and Success Stories

The safety and ease of using Epitalon are supported by scientific studies and real-life experiences. Research and reports point to Epitalon's positive health impacts without notable side effects. The lack of serious side effects holds true for both injections and oral forms of Epitalon. This flexibility in how it can be taken makes Epitalon even more appealing, allowing for personalized treatment plans.

Potential Benefits of Epitalon in Conjunction with Other Therapies

Studies of Epitalon have shown many exciting possibilities in the field of anti-aging and regenerative medicine!

Boosting Regeneration with Lifestyle Changes

An interesting part of using Epitalon is how it can be enhanced with positive lifestyle changes - especially those that reduce stress! Stress accelerates the aging process by shortening our telomeres. By combining Epitalon with stress-reducing activities such as mindfulness, acupuncture, or yoga can enhance its beneficial effects. This approach can not only help prevent stress-related damage to our cells but may also enhance the body's natural ability to regenerate.

Working Together with Traditional Treatments

Combining Epitalon with traditional treatments opens up new possibilities for improving our health. As Epitalon acts as a powerful antioxidant, similar to melatonin, it can support treatments for conditions related to oxidative stress. This antioxidant effect, combined with treatments aimed at oxidative damage, might provide a comprehensive strategy for slowing down cellular aging and disease progression. 82] Moreover, Epitalon may also be a supportive element in hormonal therapy. As it promotes increased energy levels and sleep, it could work well as a support for hormonal imbalances or the decreased hormone levels that come with aging. This combined treatment could help to improve mood, energy, and quality of life.

Enhancing Regenerative Medicine

Epitalon's potential doesn't stop there; when used with other regenerative medicine techniques, like stem cell therapies, it might offer enhanced benefits. Epitalon can boost the body's repair mechanisms, which, when used with these regenerative therapies, could lead to faster healing, better recovery, and a better outcome in anti-aging therapies. The idea is that, by making the telomeres longer, Epitalon provides a basic way for these combined treatments to work better. 83]

Supportive Measures
Alongside Epitalon Treatment

The benefits of Epitalon can be greatly increased by certain lifestyle choices and practices! For instance, a balanced diet full of essential nutrients is key to supporting the body's overall function and getting the most out of Epitalon. Eating foods that are rich in antioxidants, such as berries, nuts, greens, and dark chocolate, helps combat oxidative stress. This includes omega-3 fatty acid found in fatty fish, flaxseeds, and walnuts. This benefits heart health and helps to reduce inflammation. Both of these effects complement Epitalon's anti-aging effects. High-quality proteins from lean meats, beans, and dairy are crucial for maintaining muscle mass and our metabolic health. Consuming complex carbohydrates like whole grains, fruits, and vegetables provide a steady energy supply and are rich in fiber, vitamins, and minerals.

Water is essential for life as it supports every cellular function and ensures smooth physiological processes. Proper hydration aids in digestion, nutrient absorption, and waste elimination, facilitating an optimal internal environment for Epitalon's effectiveness. Normally, adults are encouraged to drink at least 8 glasses (about 2 liters) of water a day, although needs vary based on activity level and environmental factors.

Regular physical activity is another cornerstone of a healthy lifestyle that significantly enhances the effects of Epitalon.

It offers numerous health benefits by itself, including improved cardiovascular health, better glucose regulation, enhanced mood, and increased muscle strength. Yet, when used in combination with Epitalon these benefits are only enhanced. Incorporating a mix of aerobic exercise such as walking, cycling, or swimming, along with strength training, promotes overall well-being and amplifies the benefits of Epitalon treatment. A very basic recommendation for the average person is at least 150 minutes of moderate aerobic activity, or 75 minutes of vigorous activity, each week.

Affirming the Safety and Efficacy of Epitalon

Research and clinical observations have provided great insight into the safety and efficacy of Epitalon. For instance, a study highlighted that while Epitalon did not significantly affect body weight or overall lifespan in mice, it notably slowed down the development of spontaneous tumors and age-related changes in their biomarkers. 84] This suggests that Epitalon's impact may be more widespread, affecting our overall quality of life and health. Moreover, the safety profile of Epitalon appears very reassuring. According to studies, long-term trials with Epithalamin (of which Epitalon is a synthetic version) reported no severe adverse events in older adults over periods extending up to 12 years. 85] Such findings support the consideration of Epitalon as a very safe peptide and treatment.

Overall, all evidence available shows great promise in its safety record. While results might differ from person to person, and it is important to follow the instructions from a professional provider, the information we have now makes Epitalon a very powerful resource to enhance our vitality.

CHAPTER 14

CASE STUDIES
AND PERSONAL EXPERIENCES

Interest in Epitalon and its potential to reverse aging has been growing, both in scientific circles and among individuals focused on health. As people seek ways to live longer and slow the aging process, they're exploring a range of treatments. Epitalon has become a key player in this search. Some individuals have openly praised Epitalon, which they call the "longevity peptide," for its potential health benefits based on their personal use. They report remarkable improvements in regulating sleep, maintaining brain health, and rejuvenating their skin.

These individuals highlight that Epitalon significantly improved their sleep quality and aligned their circadian rhythm, leading to deeper, more restorative sleep. They believe these changes have not just boosted their well-being but have also positively affected their cognitive functions. Notably, some people claim that Epitalon reversed brain aging by 3.1 years and their epigenetic brain and eye age by 5.7 years, suggesting Epitalon might protect the brain and slow its aging process. Brain aging, epigenetic brain aging, and eye age are concepts that explore how our brains and eyes change as we get older.

Brain aging is the natural progression of changes in the brain's structure and function over time. This includes shrinking brain size, alterations in blood vessels, and shifts in how well we think and process information. With age, some parts of the brain might shrink, leading to slower thinking, memory issues, and challenges with multitasking.

However, not all cognitive abilities decline; some may stay the same or even improve. Brain aging is influenced by a mix of genetics, lifestyle choices, and overall health conditions.

Epigenetic brain aging involves changes in how genes work in the brain, which don't alter the DNA itself but result from environmental factors, lifestyle, and external influences. These changes can turn genes on or off, significantly affecting the brain's aging process. Epigenetics can either speed up or slow down how quickly our brains age, influencing cognitive abilities and the risk of developing age-related diseases. Researchers are looking into how these epigenetic factors play a role in aging and whether we can counteract them.

Eye age deals with how aging affects our eyes, leading to structural and functional changes that impact vision. Like the brain, the eyes experience a range of changes as we age, including less clear lenses, retina changes, and reduced sharpness of vision. These changes are behind common age-related eye problems like cataracts, macular degeneration, and glaucoma. "Eye age" also considers how the biological age of our eyes might differ from our actual age due to genetics, environmental influences, and how we live our lives.

They also point out the durability of Epitalon's effects, suggesting that it could provide lasting health and vitality benefits, not just temporary fixes. Beyond sleep and brain health, some individuals note a significant improvement in their skin's appearance, describing it as rejuvenated. This personal account supports the idea that Epitalon could have anti-aging effects, possibly by boosting collagen production and aiding skin regeneration.

Collagen is a critical protein in the skin, responsible for maintaining its structure, elasticity, and overall health. As we age, collagen production naturally decreases, leading to wrinkles, decreased skin elasticity, and other signs of aging.

Research and discussions around Epitalon suggest that it could play a significant role in combating these age-related changes in the skin. One of the key mechanisms by which Epitalon is thought to exert its effects is by stimulating skin fibroblasts. These cells are crucial for producing collagen and elastin, the proteins essential for healthy, resilient skin. By activating these fibroblasts, Epitalon may help promote the production of new collagen, potentially reversing or mitigating some of the effects of aging on the skin.

Moreover, Epitalon is believed to have antioxidant properties that further contribute to its anti-aging effects. Antioxidants play a vital role in protecting the skin from damage caused by free radicals, which can accelerate the breakdown of collagen. By combating these free radicals, Epitalon could help preserve collagen levels and maintain skin health and appearance.

Improvement of Life Quality with Epitalon

Users of Epitalon have reported several benefits, with significant enhancements in daily energy levels being a prominent advantage, leading to improved life quality. Feedback from those who have added Epitalon to their regimen indicates a substantial rise in vitality and overall health. This increase in energy corresponds with the peptide's reported effects on regulating the pineal gland, retina, and brain functions.

Some individuals have shared details about their use of Epitalon, noting key points about its administration and effects. They highlight the lack of pain after the injections, indicating a comfortable experience with Epitalon administration. Additionally, they report not feeling any burning sensation from the injections, which suggests that the treatment is well-tolerated.

This is proven by studies suggesting that any such discomfort is typically minimal or not significantly impactful compared to the potential gains from the therapy. The administration of Epitalon, when done correctly, appears to be well-tolerated.

Many users also underwent evaluations of their telomeres following the Epitalon treatment. As mentioned earlier, telomeres serve as indicators of biological aging due to their role as protective caps on chromosomes. Users discuss their telomere length in terms of "telomere years." The concept of "telomere years" refers to an assessment of an individual's biological age based on the length of their telomeres. Unlike chronological age, which is measured in years since birth, telomere years provide an insight into cellular age and, by extension, the overall health and aging pace of an individual. This measurement suggests that individuals with longer telomeres may have a biological age that is younger than their chronological age, potentially indicating better health and longevity.

Other Testimonies About Epitalon's Impact on Health and Aging

Given the diverse and significant potential benefits, individuals from various backgrounds have shared their experiences with Epitalon and provided concrete examples of how it has positively impacted their lives. People have shared valuable observations from their use of Epitalon, noting quick and distinct transformations in themselves after starting the peptide. Their comments suggest a meaningful and possibly surprising shift occurred as a result of their Epitalon regimen.

According to other testimonials, some individuals are focused on preventing their telomeres from shortening to achieve the goal of aging more slowly. They are interested in Epitalon due to its potential to maintain telomere length, a critical indicator of cellular aging, and thereby postpone the aging process.

Their decision to consider Epitalon is based on a proactive strategy to preserve health and vitality as they get older. By addressing telomere shortening, they hope to reduce the biological impacts of aging, aiming for a more youthful and energetic future.

Furthermore, people have reported experiencing positive effects from specific dosages of Epitalon, particularly highlighting its effectiveness in older individuals who might lack natural telomere extension capabilities. This indicates that Epitalon may be beneficial for those whose natural telomere maintenance processes are weakened by aging.

Health Conditions That May Benefit from Epitalon

Epitalon has gained attention in the scientific community for its potential therapeutic effects on age-related conditions, particularly age-related macular degeneration (AMD). AMD is a leading cause of vision loss among older adults and is characterized by the deterioration of the macula, a small area in the center of the retina responsible for sharp, central vision. A study highlighted in the Springer publication "Molecular Mechanisms of Retina Pathology and Ways of Its Correction" discusses the promising impact of Epitalon on retinal health. According to the research, Epitalon influenced a significant number of genes related to retinal health and showed potential as a retinoprotector for the treatment of AMD, among other retinal pathologies. [87]

The study titled "Results of the Clinical Study of Short Peptides (Cytogens) in Ophthalmology" also found that daily subcutaneous administration of Epitalon led to improvements in cases of macular damage, as evidenced by total ERG measurements. [88]

These findings suggest that Epitalon could play a role in protecting against or even reversing some of the detrimental effects of age-related macular degeneration. The peptide's potential to improve vision in patients with AMD, as indicated by small-scale studies, offers hope for those seeking alternatives to conventional treatments that often focus on slowing progression rather than reversing damage.

Furthermore, a study by Vladimir Khavinson et al. published in "Molecules" aimed to explore the effects of Epitalon on neurogenic differentiation gene expression and protein synthesis in human gingival mesenchymal stem cells (hGMSCs). The findings were remarkable, showing that Epitalon increased the synthesis of key neurogenic differentiation markers: Nestin, GAP43, β Tubulin III, and Doublecortin in hGMSCs. Furthermore, mRNA expression levels of these markers were elevated by 1.6–1.8 times in hGMSCs treated with Epitalon. 89]

The study also investigated the molecular mechanisms behind these effects, using molecular modeling methods to show that Epitalon preferably binds with specific histones (H1/3 and H1/6) at sites that interact with DNA. This interaction suggests a potential epigenetic mechanism through which Epitalon can regulate neuronal differentiation gene expression and protein synthesis in human stem cells.

Benefits of Epitalon Use Across Age Groups

Given the range of effects attributed to Epitalon, including its impact on cellular senescence, gene expression, and endocrine system regulation, it's important to know how the experience of using this compound might vary across different age groups.

Young Adults:

For younger adults, the direct anti-aging benefits of Epitalon might not be as pronounced or immediately noticeable due to their already optimal levels of cellular and endocrine function. However, studies suggest that even in younger populations, Epitalon could offer protective effects against oxidative stress and enhance the body's resistance to environmental and physiological stressors.

For example, a study highlighted by the National Center for Biotechnology Information (NCBI) found that Epitalon significantly reduced the level of reactive oxygen species (ROS) in oocytes, suggesting its potential to protect against cellular aging from an early stage. 90]

Middle-aged Adults:

Middle-aged individuals might begin to notice more substantial benefits from Epitalon usage, particularly in terms of energy levels, sleep quality, and possibly a slowdown in the visible signs of aging. This group is at a point where endocrine dysfunctions and the decline in regenerative processes start becoming more evident. Research indicates that Epitalon can restore age-related endocrine disturbances, making it a promising remedy for individuals experiencing the onset of such age-related changes. 91]

Older Adults:

One benefit particularly relevant to older adults is the improvement in sleep quality associated with Epitalon use. With age, many people experience changes in sleep patterns, including difficulty falling asleep and staying asleep. The peptide has been reported to promote deeper, more restorative sleep, which is crucial for overall health and well-being. Epitalon's broad range of biological activities also includes protective effects against age-related diseases. By inhibiting tumor development in somatic cells and potentially delaying the onset of age-related diseases such as Alzheimer's and cardiovascular diseases, Epitalon offers hope for healthier aging.

Its ability to prevent the early aging process of cells further underscores its potential as a therapeutic tool for extending a healthy lifespan. Overall, these case studies and personal experiences indicate that Epitalon could have significant benefits in anti-aging and health enhancement. These accounts from various individuals highlight the peptide's potential to positively affect longevity, improve life quality, and reduce aging effects.

Although scientific research lays the groundwork for our understanding of how Epitalon works and its possible advantages, practical experiences and testimonials are what truly demonstrate its impact on health and aging. As we look for ways to improve health span amidst the challenges of aging, Epitalon offers a promising option, suggesting a future where aging could be associated with continued vitality rather than decline. The insights gained from individuals who have experienced Epitalon therapy contribute to our knowledge and encourage ongoing investigation into this promising peptide.

CHAPTER 15

THE FUTURE OF EPITALON
IN ANTI-AGING

Through advanced computational modeling and structural biology techniques, scientists have successfully defined this peptide's molecular structure. This thorough investigation has revealed a profile characterized by remarkable stability and specificity. Vladimir Khavinson, the distinguished Russian scientist mentioned in this book, has made several groundbreaking discoveries in bioregulation and gerontology. His discovery of Epitalon significantly advanced our understanding of the aging process and created new opportunities for mitigating the effects of aging.

Khavinson's dedication to research and innovation established him as a leading figure in peptide therapy, founding the St. Petersburg Institute of Bioregulation and Gerontology. Today, this has become a key institution for conducting clinical studies on Epitalon's beneficial effects.

The Institute is now at the forefront of researching how Epitalon works, evaluating its potential as a therapy, and confirming its safety and efficacy through rigorous testing. The institute approaches its research with an interdisciplinary method that integrates molecular biology, biochemistry, pharmacology, and clinical medicine, creating a vibrant and creative scientific environment. The main goal is to explore the molecular workings of Epitalon's anti-aging effects.

This involves a combination of laboratory and animal studies aimed at understanding how Epitalon affects key biological processes related to aging, such as managing the length of telomeres, repairing DNA, and delaying cell senescence, as covered in this book. Extensive clinical trials have been conducted to determine Epitalon's effectiveness in treating a wide range of conditions typically associated with aging.

Epitalon's Effect On Histones

Histone proteins, located in the nuclei of eukaryotic cells, are essential for the organization and packaging of DNA into nucleosomes. Eukaryotic cells are characterized by their complex structure, distinguished by the presence of a nucleus enclosed within a membrane. These cells form the foundational building blocks of a wide range of organisms, such as animals, plants, fungi, and protists (not animals, plants, or fungi). Nucleosomes, on the other hand, are the structural units forming chromatin in chromosomes. These units consist of DNA strands coiled around histone octamers.

Histones are classified into five primary types: H1, H2A, H2B, H3, and H4, each with unique structural attributes and roles within the nucleosome. The H1 histone, often referred to as the linker histone, attaches to DNA segments between nucleosomes, aiding in the stabilization of larger chromatin structures. Meanwhile, H2A, H2B, H3, and H4 collaborate to form the core of the octamer that DNA wraps around within the nucleosome.

Histones are pivotal for several vital processes in the organism:

- DNA Packaging:
  They compact DNA into chromatin, efficiently condensing lengthy DNA molecules to fit within the limited nucleus space. This compaction is vital for gene expression regulation and the genome's structural integrity.

- Gene Regulation:
  Histone modifications, including acetylation, methylation, phosphorylation, and ubiquitination, influence chromatin structure to either enhance or restrict DNA accessibility to transcription factors and other proteins, thereby managing gene activity.

- Epigenetic Inheritance:
  Modifications to histones can be transmitted from parent to offspring cells during division, effectuating the lineage of epigenetic characteristics without altering the DNA sequence. This inheritance is fundamental to cell differentiation, development, and identity preservation.

- DNA Repair:
  Histones are instrumental in DNA repair by drawing repair proteins to damaged DNA sites and supporting their function. Adequate histone modification and chromatin remodeling are critical to effective DNA repair and genomic stability.

- Chromosome Segregation:
  In cell division, histones assist in chromosome organization and condensation, ensuring the precise distribution of genetic material to daughter cells. Histone modifications and chromatin remodeling are crucial in controlling these processes.

Histones are integral to the functions of the nucleus, facilitating everything from DNA packing and gene regulation to epigenetic heredity, DNA repair, and chromosome segregation.

Their interaction with DNA and nuclear proteins is key to gene expression regulation and the essential cellular activities that support the growth, operation, and preservation of multicellular organisms. AEDG, or Epitalon, has been noted for its specific interactions with histones, which are key in maintaining chromatin structure and influencing gene regulation within the nucleus. Chromatin is composed of DNA, RNA, and proteins found within the nucleus of eukaryotic cells. In addition to the nucleus, eukaryotic cells are equipped with various membrane-bound organelles like mitochondria, endoplasmic reticulum, and Golgi apparatus, which perform specialized functions that are crucial for the cell's survival and operation.

Chromatin's primary function is to package the long strands of DNA into a more compact, denser shape, allowing them to fit inside the cell nucleus. This configuration not only provides structural support to the DNA but also plays a crucial role in regulating gene expression and DNA replication. By controlling the accessibility of certain parts of the DNA to the cellular machinery, chromatin ensures that genes are turned on or off at the appropriate times. Vladimir Khavinson and his colleagues' study from 2020 showed that Epitalon has a significant interaction with histones. Histones have been shown to interact with Epitalon, especially with histone variants such as H1/1, H1/3, H1/6, H2b, H3, and H4.

These interactions show the probable mechanisms by which Epitalon affects many biological processes! As mentioned, Epitalon is a synthetic peptide composed of the amino acids alanine (A), glutamic acid (E), aspartic acid (D), and glycine (G). The unique combination of these amino acids affects its hydrophobicity and structural characteristics. "Hydrophobicity" refers to the inclination of the peptide's amino acid residues to repel water. This influences the peptide's solubility, structure, folding, and interactions within biological systems. Hydrophobic amino acids tend to avoid water and, as a result, often locate themselves within the core of proteins or peptides, away from the aqueous environment.

This trait influences the peptide's ability to bind to and interact with other hydrophobic molecules or molecular structures, such as membrane lipids. Consequently, the hydrophobic property of a peptide like Epitalon can significantly impact its effectiveness and mechanism of action in biological processes, including its specific interactions with the hydrophobic cores of histones within chromatin. The amino acid Alanine, which is hydrophobic with a nonpolar side chain, contrasts with glutamic and aspartic acids, which are hydrophilic (the affinity of a molecule or substance to mix with or dissolve in water) due to their negatively charged polar side chains. Additionally, the amino acid glycine has a non-polar side chain.

Epitalon has been shown to have selective binding patterns to different histone variants, attaching to H1/1, H1/3, and H1/6 at various sites with differing affinities and energies. It similarly engages with H2b, H3, and H4 histones, mainly targeting certain areas like the N-terminal domain or the "tail" region. The connection of Epitalon with the histone N-terminal domains, particularly of H3 and H4, points to possible pathways through which Epitalon may alter chromatin architecture and gene activity! 92]

Such interactions might influence histone modifications, the remodeling of chromatin, and the transcription of genes, potentially affecting cellular processes related to aging, DNA restoration, and cellular longevity. This information offers us crucial insight into its action at the molecular level and showcases its potential uses in many forms of therapy and gene expression therapies.

Biotechnological Innovations and Epitalon's Accessibility

This discovery regarding Epitalon's structure holds great promise for its potential large-scale production. Recent advancements in biotechnology and peptide synthesis techniques have made this formerly niche compound much more accessible, and we now see it widely available within aesthetic and athletic products. Defining the molecular structure of Epitalon was a significant milestone in biotechnological research.

This development enabled the mass production of Epitalon and influenced peptide synthesis and biotechnology. Innovations like recombinant DNA technology have been key in the industrial production of peptides, including Epitalon. By employing genetic engineering, scientists can modify organisms like bacteria or yeast to produce large quantities of peptides efficiently and economically. This method provides a scalable and consistent approach to peptide synthesis that surpasses previous methods. Specifically, using bacterial expression systems shows how genetic modification of bacterial hosts, such as Escherichia coli (E. coli), can produce and secrete recombinant peptides in large volumes.

According to Bloom Tech, the creation of Epitalon is categorized into:

1. chemical synthesis,
2. and biosynthesis

Chemical Synthesis

Epitalon is represented by the molecular structure C14H22N4O9. In the chemical synthesis approach, synthesizing Epitalon includes preparing the reactants, which are alanine (Ala), glutamic acid (Glu), asparagine (Asp), and lysine (Lys), alongside acylating agents like Boc-Lys-OtBu and Asp(OtBu)2. This ensures that these components are over 99% pure.

The synthesis process unfolds through several stages:

The creation of alanine-4-hydroxybutyric anhydride (Ala-Hyp) first begins by mixing alanine (Ala) with 4-hydroxybutyric anhydride (Hyp-OtBu).

Then, an acylation reaction is initiated using activators such as DCC or EDC in an anhydrous setting to produce alanine-4-hydroxybutyric anhydride (Ala-Hyp). This results in white crystals with a purity of over 95%.

The assembly of Ala-Hyp-Glu-OtBu involves combining the previously synthesized alanine-4-hydroxybutyric anhydride with glutamic acid butyrate (Glu-OtBu), followed by a series of condensation reactions in an anhydrous environment to achieve Ala-Hyp-Glu-OtBu.

The final product is a white powder with a purity above 95%. To synthesize Epitalon, compounds such as Asp(OtBu)2 and Boc-Lys-OtBu are incorporated into the condensation reaction system. This is done in a pre-planned sequence, and proceeds through multiple condensation stages, including

A Deprotection Phase

Deprotecting Asp(OtBu)2 is done by removing its protective group using sodium hydroxide (NaOH) and trichloroacetic acid (TCA). This converts it into Asp units while simultaneously releasing BuOt. The reaction occurs at room temperature for about an hour.

Following the reaction, an acid-base neutralization is performed, complemented by the addition of an ample amount of saturated sodium chloride solution. Then ethanol precipitation and vacuum drying are done to obtain Asp as a white solid.

The Condensation Phase

By adding Ala-Hyp-Glu-OtBu and Asp into the system, the condensation process is executed to form Epitalon. First, the Ala-Hyp protecting group is removed by dissolving Ala-Hyp-Glu-OtBu in methanol, with trichloroacetic acid (TCA) and water added. The reaction is done at room temperature, with subsequent NaOH treatment to neutralize acidity. This continues with the removal of the Glu-OtBu and Lys protecting groups, ultimately producing Epitalon.

The reaction is then followed by an assessment to confirm the product's characteristics and purity, using analytical methods defined by the European Pharmacopoeia (EP) or the United States Pharmacopeia (USP).

Biosynthesis

Biosynthesis involves leveraging microorganisms or synthetic enzymes through fermentation and enzyme catalysis approaches. This includes:

The fermentation technique genetically modifies Escherichia coli to produce Epitalon. The first step is to select a suitable host bacteria for the expression. The Epitalon gene sequence is introduced into the selected host via DNA recombination technology. After transformation and cultivation, the host synthesizes Epitalon, which is then isolated using various purification techniques to achieve a high-purity product.

The enzyme-catalyzed method connects different amino acids with specific enzymes to formulate Epitalon. This can include using L-glutamate-5-aminase for the synthesis of Glu-OtBu from glutamate and butyrate, followed by the use of L-asparaginase for the condensation of asparagine and Ala-Hyp-Glu-OtBu to create Epitalon.

By leveraging the metabolic capabilities of microorganisms like bacteria or yeast, fermentation procedures can be optimized, offering benefits in terms of cost, scalability, and even environmental sustainability. This can make it a superior choice for commercial peptide production. The improved production and availability of Epitalon also provide researchers with sufficient quantities of Epitalon for clinical studies.

Epitalon's synthesis is a sophisticated process, and biosynthesis offers a promising technique that can still be further developed. Its potential medical and health-promoting applications relate to treatments for anti-aging, enhancing our immune system, for neuroprotection, and even cancer, as discussed throughout this book.

Solid-Phase Peptide Synthesis

Additionally, improvements in solid-phase peptide synthesis (SPPS) techniques have improved the efficiency and scalability of Epitalon production. SPPS allows for the sequential addition of amino acids on a solid support to assemble the desired peptide and has been refined through advancements in automation and new resin materials. These enhancements have increased the yield and purity of Epitalon, making it a more economical option for large-scale manufacturing.

Epitalon as a Future Adjuvant to Cancer Therapy

Cancer has become one of the most significant health challenges of our time. Today, a complex array of cancer forms have been identified and connected with degenerative aging. For instance, pancreatic cancer is notably aggressive and prone to early metastasis. This frequently results in diagnosis first being done at a late stage, which complicates the effectiveness of treatments. Glioblastoma multiforme, a certain form of brain cancer, is distinguished by its vigorous growth and the difficulty of treatment due to its tendency to infiltrate brain tissues, and the lack of success of current available therapies.

When it comes to ovarian cancer, this is often only detected once it has progressed beyond the ovaries, often diminishing the survival rate.

Its subtype, "triple-negative breast cancer," is characterized by the absence of estrogen and progesterone receptors, and no HER2 protein expression. This typically makes it unresponsive to hormone treatment and targeted therapies.

200

Furthermore, metastatic melanoma (when spread to other organs) is significantly hard to treat due to its aggressive nature. Lung cancer, especially advanced non-small cell lung cancer (NSCLC), often builds up resistance to chemotherapy and targeted treatments, greatly minimizing the efficacy of conventional treatments.

These types of cancer are particularly challenging to manage and treat because of their overall aggressive and accelerated nature. Conventional treatment itself also comes with its own host of damaging side effects, some of which are permanent, like neuropathy and vascular injury (damage to the blood vessels).

A study was conducted on female transgenic FVB mice, carrying the breast cancer gene HER-2/neu. The mice were injected monthly with either Vilone or Epithalon, where Epithalon was administered subcutaneously at a dose of 1 microgram for 5 consecutive days, starting from the 2nd month of life. The study found that Epithalon significantly inhibited the development of breast neoplasms in the mice compared to the control group. Specifically, the maximum size of breast adenocarcinomas in the Epithalon-treated group was 33% lower than in the control group. Additionally, the intensity of HER-2/neu mRNA expression in breast tumors of the Epithalon-treated mice was found to be 3.7 times lower than in control animals.

This suggests that Epithalon has inhibitory effects on breast tumor development in transgenic mice, possibly through the suppression of HER-2/neu expression. 93]

Another study examined the effect of Epitalon on colon carcinogenesis in rats. Eighty 2-month-old outbred male LIO rats were divided into four groups and exposed to weekly subcutaneous injections of 1,2-dimethylhydrazine (DMH) at a single dose of 21 mg/kg body weight. Some of the rats received subcutaneous injections of saline as a control (group 1), while others were injected with Epitalon at a dose of 1 microgram. This was done either throughout the entire experiment (group 2), after termination of carcinogen injections (group 3), or during the period of DMH exposure (group 4).

Colon carcinomas developed in 90-100% of the DMH-treated rats. The study found that the number of total colon tumors per rat was significantly lower in groups 2 and 4, compared to the control group (group 1). In the rats from group 2, colon tumors were also found to be smaller than in control animals. Additionally, the incidence and multiplicity of tumors in the ascending and descending colon were significantly decreased in group 2, compared to group 1. Group 4 also showed a significant decrease in the mean number of tumors per rat.

Additionally, the study saw a trend towards decreased tumor numbers in the rectum in rats treated with Epitalon in groups 2, 3, and 4. Epitalon was also found to inhibit the development of tumors in the jejunum and ileum. The jejunum is the middle section of the small intestine, located between the duodenum (the first part) and the ileum (the last part). It is key in absorbing nutrients, like carbohydrates, proteins, and fats, into the bloodstream. The ileum is the final part of the small intestine, connecting to the large intestine (colon) at the ileocecal valve. It absorbs nutrients that were not absorbed by the jejunum, as well as bile salts and vitamin B12.

Both the jejunum and ileum have extensive surface areas lined with villi and microvilli, which increase their absorptive capacity.

Overall, the results of the study indicate that Epitalon has an inhibitory effect on chemically induced bowel carcinogenesis in rats. 94]

As Epitalon has been shown to not only directly target the disease but also enhance the effectiveness of conventional treatments, its capacity to regulate crucial signaling pathways within cancer cells indicates a potential alternative treatment. Additionally, it could possibly contribute to a reduction in the harmful side effects associated with many traditional cancer therapies.

Neurodegeneration And Cognitive Enhancement

Age-related cognitive decline and neurodegenerative diseases are another significant challenge our healthcare system faces. Several neurodegenerative conditions, such as Alzheimer's, Parkinson's, and amyotrophic lateral sclerosis (ALS), are very difficult to manage due to the absence of an identified, definite cause. This lack of insight into root mechanisms makes developing any successful treatment extremely difficult. Currently, none of these diseases can be cured, and available treatments mainly focus on symptom management with the intention of slowing the progression of the disease.

The Blood-Brain Barrier

One of the main obstacles in treating these conditions is the blood-brain barrier, or BBB. This highly selective barrier maintains a controlled environment for the brain by regulating which substances can pass from the bloodstream into the brain.

The BBB is essential for protecting the brain from potential harm; however, it significantly complicates the delivery of drugs to the brain. The challenge of delivering therapeutic agents across the BBB is compounded by its selective nature. Furthermore, as the BBB tends to become less restrictive as we age 95], this naturally complicates the risk of infections and problematic issues.

To address these drug delivery issues, many innovative approaches are being developed. Examples of these are the use of nanoparticles, liposomes, and viral vectors. These methods are aimed at either bypassing the BBB's barriers or facilitating the entry of therapeutic substances into the brain to maximize their effectiveness while limiting unintended side effects.

A study in 2006 explored the characteristics and impacts of Epitalon on rat motor cortex spontaneous activity. By using intranasal delivery, as a non-invasive method to circumvent the blood-brain barrier, researchers wanted to determine Epitalon's effects. Male Wistar rats were sedated using urethane, and their motor cortex's spontaneous activity was monitored through extracellular recordings from glass microelectrodes. Initially, motor cortex activity was recorded for 10-15 minutes. After this baseline period, Epitalon was administered intranasally at a 2 nanogram dose, with the recording extending for an additional 30 minutes.

Results indicated a marked increase in neural activity soon after Epitalon had been administered - observing a 2 to 2.5 times rise in the frequency of neuronal spikes! The rats' reaction to Epitalon featured several stages, with the first surge in activity noted 5 to 7 minutes after the infusion. This heightened activity continued through the second (11-12 minutes) and third phases (17-18 minutes), showing a prolonged effect of Epitalon on neural activity. The initial increase in neuronal activity was linked directly to Epitalon's effect on motor cortex cells, enhancing the activity of already active units, and activating previously inactive ones. This was found to boost spontaneous neural activity overall. 96]

Another study by Khavinson examined how the peptides Epitalon and Vilone affect melatonin production in rat pinealocyte cultures, specifically focusing on their impact on the arylalkylamine-N-acetyltransferase (AANAT) enzyme and the pCREB transcription protein (key in producing melatonin). The research found that Epitalon boosts the levels of AANAT and pCREB, as well as the amount of melatonin present in the culture medium, showcasing a beneficial effect on the pathway responsible for creating melatonin in pinealocytes. Pinealocytes are specialized cells found within the pineal gland, located in the brain.

These cells are primarily responsible for the production of melatonin. Melatonin secretion is influenced by light exposure to the eyes, with pinealocytes activating at night to increase melatonin levels, thereby promoting sleep, and reducing production during the day. The study also observed that adding norepinephrine, a neurotransmitter involved in controlling melatonin, along with the peptides, led to a higher expression of AANAT and pCREB. This suggests that norepinephrine and the peptides work together to enhance the production of essential enzymes and transcription factors needed for making melatonin! The findings suggest that Epitalon and Vilone peptides can play a role in influencing melatonin production in pinealocytes, possibly by activating AANAT and pCREB. 97]

Understanding the peptides' ability to influence melatonin production raises some interesting questions about their potential cognitive effects. Melatonin is known not only for regulating our sleep cycles, but also for its neuroprotective properties, and how these impact our cognitive health. If Epitalon and Vilone peptides facilitate increased melatonin production, they could indirectly support cognitive functions by enhancing sleep quality, and neuroprotection.

Several other findings from both preclinical studies and human trials indicate that Epitalon has neuroprotective properties. Combined with its ability to improve cognitive function, this could transform the treatment and management of conditions like Alzheimer's and dementia.

These findings offer more than just speculation. They show us a glimpse of a potential future where age-related neurodegeneration might no longer be a threat to our health.

Inspiring Innovative
Drug Delivery Systems

The distinctive structural design of Epitalon as a short peptide has been widely examined for its therapeutic potential in new drug developments. Could Epitalon inspire the development of the next generation of drug delivery systems?

An extensive review was done on short peptides, which found that they exhibit remarkable qualities! They are cost-effective to produce, have strong mechanical properties, can easily penetrate tissues, and are less likely to trigger immune reactions compared to longer peptides. They can be easily managed, and modified, and are compatible with biological tissues, and cells, without causing damage or toxic effects. They are also biodegradable, and generate a minimal degree of immune response. The ability of these peptides to support the growth, and change, of various cell types simplifies the process of refining structures, and their small size makes them particularly suitable for oral delivery methods.

Many of them are crucial in forming precise nanostructures that can, for example, turn into hydrogels with antimicrobial properties, speeding up wound healing. Short peptides can also overcome the problem of short-lived drug effects by allowing for controlled release, which is especially significant in treatments directed at brain tissue repair, and in managing neurodegenerative diseases. Their broad applicability extends to nano-theranostic (therapeutic and diagnostic) approaches, for instance in inhibiting cancer cell growth, and in developing advanced drug formulations that enhance treatment effectiveness and safety. 98]

Short peptides play a crucial role in researching neurodegenerative diseases. Understanding how they form amyloid-like structures can help us get a better grasp of amyloid fibrils' formation, and their properties. "Amyloid" refers to aggregates of proteins that become folded into a shape that allows many copies of that protein to stick together, forming fibrils. These protein accumulations can be found in several organs and tissues, disrupting their normal function. The presence of amyloid deposits is associated with various diseases, including Alzheimer's disease, and other neurodegenerative conditions, where they can impair cognitive functions by affecting brain tissue.

These short peptides, including those that can penetrate cells, are vital for developing effective drug delivery systems, targeting cells, and preventing amyloid β aggregation. An exciting new area of research is found in using peptide-based gels to encapsulate drugs. Peptide-based drugs have been shown to promote neuronal growth, encourage neurogenesis, and improve memory – all while offering a personalized approach to therapy! 98]

Many short peptides, including Epitalon, play a crucial role in a wide range of biological functions. This alone makes them very valuable for innovative treatments across various fields. Their use is now notable in biomedicine, diagnostics, pharmaceuticals, and cosmeceuticals. Both natural and synthetic short peptides are now being explored for their applications in nanotechnology, supported by advancements in structural bioinformatics and supramolecular chemistry (the study of structures formed by molecules interacting through non-covalent bonds, such as hydrogen bonding).

The combination of in silico techniques and sophisticated synthesis methods is creating new ways of developing peptide-based compounds, with wide-ranging uses.

Nanoparticle formulations designed for precise targeting, or biomimetic carriers inspired by its molecular mechanisms, have the potential to enhance Epitalon's efficacy significantly. In this context, Epitalon may not only be a therapeutic agent but also a vital part of future personalized medicine.

Extending Beyond
the Realm of Anti-Aging

The growing interest in peptide therapy throughout recent years has opened up new avenues of research and not just within the field of aesthetic and cosmetological enhancement. It now also significantly extends into the strengthening of our vitality, innovations in personalized medicine, and the proactive prevention of what were previously considered incurable diseases.

These peptides represent a key in unlocking the potential for not only living longer but ensuring that those additional years are marked by true health and vitality. The focus is now shifting towards an integrative approach where the objective is not merely to extend life but enrich its quality, tailoring care to our individual genetic makeup, and preemptively addressing health issues before they manifest. This perspective on longevity emphasizes a future where aging is not synonymous with decline but with sustained well-being and resilience. To me, this epitomizes what should be the ultimate goal of modern medicine and science!

We stand on the brink of a new era — a world where humanity can progress with confidence, grounded in scientific understanding. While the promise of Epitalon is still evolving, it represents a profound testament to our human curiosity and our relentless pursuit of knowledge and health.

CHAPTER 16

EPILOGUE: THE FUTURE OF LONGEVITY AND ANTI-AGING THERAPY

Dear reader, as we reach the conclusion of our discussion on the incredible benefits of Epitalon, I want to reflect on what we've covered and consider the exciting future in the field of longevity and anti-aging therapy.

Throughout this book, we've seen how Epitalon influences cellular health, particularly its role in activating telomerase and maintaining telomere length. This mechanism holds the potential to delay cellular aging, offering the possibility to significantly slow or even reverse age-related decline. Understanding and leveraging this process could be a cornerstone in future anti-aging strategies, potentially extending healthy lifespans in ways we are just beginning to understand.

Building on this foundation, we examined Epitalon's impact on metabolic and endocrine systems. By supporting hormonal balance and enhancing metabolic processes, Epitalon addresses critical aspects of aging such as muscle loss and hormonal imbalances. These benefits highlight its potential to help maintain vitality, fitness, and strength as we age. The implications for improving quality of life are profound, suggesting that we can age not only longer but healthier and more robustly.

Quality sleep is another essential component of longevity, and Epitalon's ability to regulate circadian rhythms and enhance melatonin production underscores its importance. Improved sleep quality not only rejuvenates our bodies but also supports cognitive function, emotional health, and immune response.

This holistic approach to enhancing sleep offers a practical solution to one of the most common age-related challenges. Imagine a future where sleep disorders are significantly reduced, and restful, restorative sleep is the norm for all ages.

Our skin, the body's largest organ, benefits significantly from Epitalon. By promoting collagen synthesis and protecting against oxidative stress, Epitalon helps maintain skin elasticity, firmness, and resilience. This natural approach to skincare can effectively counteract visible signs of aging, supporting a youthful appearance and overall skin health. It's not just about looking younger; it's about maintaining the skin's functional integrity as a barrier and a detoxification organ.

Epitalon's antioxidative properties also extend to eye health, protecting vision and preventing age-related ocular conditions. Maintaining optimal eye health is crucial for preserving quality of life and independence as we age, making this peptide's potential in eye care particularly noteworthy. Clear vision is integral to enjoying life fully, and interventions like Epitalon could significantly reduce the prevalence of debilitating eye conditions.

Bone and joint health, often compromised by conditions like osteoporosis and osteoarthritis, can also be supported by Epitalon. Enhancing bone density and promoting joint health are essential for maintaining mobility and reducing the risk of fractures. This regenerative potential underscores the importance of Epitalon in supporting the skeletal system. Imagine a future where the pain and immobility associated with aging joints are significantly mitigated, allowing for a more active and engaged lifestyle.

As we consider the broad spectrum of benefits Epitalon offers, it's clear that this peptide does not target one aspect of aging. Instead, it presents a multi-faceted approach to enhancing our overall health and vitality as we age more gracefully. All the benefits combined make a comprehensive strategy for promoting longevity as never seen before!

The potential applications of Epitalon offer a previously unseen promising future in longevity and anti-aging therapy. Its ability to address various aspects of aging — from cellular health to metabolic balance, from sleep regulation to skin vitality, from eye protection to skeletal strength — paints a comprehensive picture of what the future may hold. This peptide represents a shift towards more holistic and integrative health solutions, moving beyond symptomatic treatment to address underlying causes of aging.

As we move forward, it is essential to continue supporting and investing in scientific research to fully understand and harness the potential of Epitalon and other groundbreaking therapies. The pursuit of longevity is not merely about extending our years but enhancing the quality of those years, ensuring that we age with grace, vitality, and dignity. The research community plays a crucial role in this, and our collective support can drive these innovations from the lab to real-world applications.

Thank you for taking the time to read this book and gain insight into the possibilities that Epitalon might offer you. I hope this book has inspired you to consider how these advancements can be integrated into your approach to health and wellness.

To me, the future of longevity looks bright! Together, we can create a much healthier, more youthful tomorrow! Your engagement and interest in your own health are vital in making this vision a reality.

Yours in health and longevity,

Dr. Jon Harmon, DC

REFERENCES

FROM AN INTRODUCTION TO EPITALON
– CHAPTER 1:

1] Source: Anisimov, V. N., Khavinson, V. K., Popovich, I. G., Zabezhinski, M. A., Alimova, I. N., Rosenfeld, S. V., ... & Yashin, A. I. (2003). Effect of Epitalon on biomarkers of aging, life span and spontaneous tumor incidence in female Swiss-derived SHR mice. Biogerontology, 4(4), 193-202.

2] Source: Khavinson, V.Kh., Izmaylov, D.M., Obukhova, L.K., & Malinin, V.V. (2000). Effect of epitalon on the lifespan increase in Drosophila melanogaster. Mechanisms of Ageing and Development, 120(1-3), 141-149.

FROM THE SCIENCE OF TELOMERES
– CHAPTER 2:

3] Shi H, Li X, Yu H, Shi W, Lin Y, Zhou Y. Potential effect of dietary zinc intake on telomere length: A cross-sectional study of US adults. Front Nutr. 2022 Nov 16;9:993425. doi: 10.3389/fnut.2022.993425. PMID: 36466397; PMCID: PMC9709254.

4] Adli A, Hosseini SM, Lari Najafi M, Behmanesh M, Ghezi E, Rasti M, Kazemi AA, Rad A, Falanji F, Mohammadzadeh M, Miri M, Dadvand P. Polycyclic aromatic hydrocarbons exposures and telomere length: A cross-sectional study on preschool children. Environ

Res. 2021 Apr;195:110757. doi:
10.1016/j.envres.2021.110757. Epub 2021 Jan 23.
PMID: 33493537.

FROM EPITALON AND TELOMERE ACTIVATION – CHAPTER 3:

4] Khavinson, V.K., Bondarev, I.E., & Butyugov, A.A.
Epithalon Peptide Induces Telomerase Activity and
Telomere Elongation in Human Somatic Cells. Bulletin
of Experimental Biology and Medicine 135, 590–592
(2003). https://doi.org/10.1023/A:1025493705728

5] Source: Goncharova N.D., Lapin B.A., Khavinson
V.Kh. (2002). Age-Associated Endocrine Dysfunctions
and Approaches to Their Correction. Bulletin of
Experimental Biology and Medicine, 134(5), 417-421.

6] Ilina, A., Khavinson, V., Linkova, N., & Petukhov, M.
(2022). Neuroepigenetic Mechanisms of Action of
Ultrashort Peptides in Alzheimer's Disease. International
Journal of Molecular Sciences, 23(4259.
https://doi.org/10.3390/ijms23084259

FROM CLINICAL STUDIES ON EPITALON – CHAPTER 4:

7] Zamyatnin, A.A. Gerontological Aspects of Genome
Peptide Regulation (by V. Kh. Khavinson and V. V.
Malinin, KARGER, 2005, 104 p., 70€). Biochemistry
(Moscow) 70, 1185 (2005).
https://doi.org/10.1007/s10541-005-0245-6
DOI:10.1007/s10541-005-0245-6

8] Khavinson, V.; Diomede, F.; Mironova, E.; Linkova, N.; Trofimova, S.; Trubiani, O.; Caputi, S.; Sinjari, B. AEDG Peptide (Epitalon) Stimulates Gene Expression and Protein Synthesis during Neurogenesis: Possible Epigenetic Mechanism. Molecules 2020, 25, 609. https://doi.org/10.3390/molecules25030609

9] Anisimov VN, Khavinson VK, Provinciali M, Alimova IN, Baturin DA, Popovich IG, Zabezhinski MA, Imyanitov EN, Mancini R, Franceschi C. Inhibitory effect of the peptide epitalon on the development of spontaneous mammary tumors in HER-2/neu transgenic mice. Int J Cancer. 2002 Sep 1;101(1):7-10. doi: 10.1002/ijc.10570. PMID: 12209581.

10] Khavinson VKh, Tarnovskaya SI, Linkova NS, Pronyaeva VE, Shataeva LK, Yakutseni PP. Short cell-penetrating peptides: a model of interactions with gene promoter sites. Bull Exp Biol Med. 2013 Jan;154(3):403-10. doi: 10.1007/s10517-013-1961-3. PMID: 23484211.

11] Khavinson, V.K. Peptides, genome, aging. Adv Gerontol 4, 337–345 (2014). https://doi.org/10.1134/S2079057014040134

12] Khavinson, & Diomede, & Mironova, Ekaterina & Linkova, & Trofimova, & Trubiani, Oriana & Caputi, Sergio & Sinjari, Bruna. (2020). AEDG Peptide (Epitalon) Stimulates Gene Expression and Protein Synthesis during Neurogenesis: Possible Epigenetic Mechanism. Molecules. 25. 609. 10.3390/molecules25030609.

13] Anisimov VN, Khavinson VKh, Popovich IG, Zabezhinski MA, Alimova IN, Rosenfeld SV, Zavarzina NY, Semenchenko AV, Yashin AI. Effect of Epitalon on biomarkers of aging, life span and spontaneous tumor incidence in female Swiss-derived SHR mice. Biogerontology. 2003;4(4):193-202. doi: 10.1023/a:1025114230714. PMID: 14501183.

14] Rosenfeld SV, Togo EF, Mikheev VS, Popovich IG, Khavinson VKh, Anisimov VN. Effect of Epitalon on the incidence of chromosome aberrations in senescence-accelerated mice. Bull Exp Biol Med. 2002 Mar;133(3):274-6. doi: 10.1023/a:1015899003974. PMID: 12360351.

15] Khavinson VKh, Lin'kova NS. [Morphofunctional and molecular bases of pineal gland aging]. Fiziol Cheloveka

FROM CARDIOVASCULAR BENEFITS – CHAPTER 5:

16] Avolio F, Martinotti S, Khavinson VK, Esposito JE, Giambuzzi G, Marino A, Mironova E, Pulcini R, Robuffo I, Bologna G, Simeone P, Lanuti P, Guarnieri S, Trofimova S, Procopio AD, Toniato E. Peptides Regulating Proliferative Activity and Inflammatory Pathways in the Monocyte/Macrophage THP-1 Cell Line. Int J Mol Sci. 2022 Mar 25;23(7):3607. doi: 10.3390/ijms23073607. PMID: 35408963; PMCID: PMC8999041.

17] "Medical Professional Monograph Epithalon," by the International Peptide Society, no year or author noted. https://peptidesociety.org/wp-content/uploads/2018/07/Epithalon-Monograph-Final.pdf

18] Anisimov VN, Arutjunyan AV, Khavinson VK. Effects of pineal peptide preparation Epithalamin on free-radical processes in humans and animals. Neuro Endocrinol Lett. 2001;22(1):9-18. PMID: 11335874.

19] Source: Zhan, Y., & Hägg, S. (2019). Telomere length and cardiovascular disease risk. Current Opinion in Cardiology, 34(3), 270-274.https://doi.org/10.1097/HCO.0000000000000613

20] Source: Cherkashin, V. A., Semin, G. F., & Veretenko, A. A. (2002). Optimization of cardiovascular function by peptide bio-regulators.

21] Source: Korkushko, O. V., Khavinson, V. Kh., Shatilo, V. B., & Antonyuk-Shcheglova, I. A. (2006). Geroprotective effect of Epithalamine (pineal gland peptide preparation) in elderly subjects with accelerated aging. Bull Exp Biol Med, 142(3), 356-9. doi: 10.1007/s10517-006-0365-z

FROM NEUROLOGICAL PROPERTIES – CHAPTER 6:

22] Promoting Brain Health: Be a champion! Make a difference today!, Center for Disease Control and Prevention, funded by the U.S. Department of Health and Human Services, www.cdc.gov, https://www.cdc.gov/aging/pdf/cognitive_impairment/cogimp_genaud_final.pdf

23] CNN, article "Stress may lead to lower cognitive function, study finds," by Deidre McPhillips, Updated 1:23 PM EST, Tue March 7, 2023, https://www.cnn.com/2023/03/07/health/high-stress-lower-cognition-study-wellness/index.html#:~:text=CNN%20%E2%80%94-,People%20with%20elevated%20stress%20levels%20are%20more%20likely%20to%20experience,poor%20immune%20response%20and%20more.

24] Alzheimer's Association, article What Is Alzheimer's?, © 2024, no author, https://www.alz.org/alzheimers-dementia/what-is-alzheimers

25] Source: Khavinson, V., Diomede, F., Mironova, E., Linkova, N., Trofimova, S., Trubiani, O., Caputi, S., & Sinjari, B. (2020). AEDG Peptide (Epitalon) Stimulates Gene Expression and Protein Synthesis during Neurogenesis: Possible Epigenetic Mechanism. Molecules, 25(3), 609. https://www.mdpi.com/1420-3049/25/3/609

26] Source: Ilina, A., Khavinson, V., Linkova, N., & Petukhov, M. (2022). Neuroepigenetic Mechanisms of Action of Ultrashort Peptides in Alzheimer's Disease. International Journal of Molecular Sciences, 23(8), 4259. https://www.mdpi.com/1422-0067/23/8/4259

27] Khavinson VKh, Lezhava TA, Monaselidze JR, Jokhadze TA, Dvalishvili NA, Bablishvili NK, Trofimova SV. Peptide Epitalon activates chromatin at the old age. Neuro Endocrinol Lett. 2003 Oct;24(5):329-33. PMID: 14647006.

28] Khavinson V, Diomede F, Mironova E, Linkova N, Trofimova S, Trubiani O, Caputi S, Sinjari B. AEDG Peptide (Epitalon) Stimulates Gene Expression and Protein Synthesis during Neurogenesis: Possible Epigenetic Mechanism. Molecules. 2020 Jan 30;25(3):609. doi: 10.3390/molecules25030609

FROM IMMUNE SYSTEM FORTIFICATION – CHAPTER 7:

29] Mannick JB, Morris M, Hockey HP, Roma G, Beibel M, Kulmatycki K, Watkins M, Shavlakadze T, Zhou W, Quinn D, Glass DJ, Klickstein LB. TORC1 inhibition enhances immune function and reduces infections in the elderly. Sci Transl Med. 2018 Jul 11;10(449):eaaq1564. doi: 10.1126/scitranslmed.aaq1564. PMID: 29997249.

30] Lin'kova NS, Kuznik BI, Khavinson VKh. [Peptide Ala-Glu-Asp-Gly and interferon gamma: their role in immune response during aging]. Adv Gerontol. 2012;25(3):478-82. Russian. PMID: 23289226.

31] Avolio F, Martinotti S, Khavinson VK, Esposito JE, Giambuzzi G, Marino A, Mironova E, Pulcini R, Robuffo I, Bologna G, Simeone P, Lanuti P, Guarnieri S, Trofimova S, Procopio AD, Toniato E. Peptides Regulating Proliferative Activity and Inflammatory Pathways in the Monocyte/Macrophage THP-1 Cell Line. Int J Mol Sci. 2022 Mar 25;23(7):3607. doi: 10.3390/ijms23073607. PMID: 35408963; PMCID: PMC8999041.

32] Kazakova TB, Barabanova SV, Khavinson VKh, Glushikhina MS, Parkhomenko EP, Malinin VV, Korneva EA. In vitro effect of short peptides on expression of interleukin-2 gene in splenocytes. Bull Exp Biol Med. 2002 Jun;133(6):614-6. doi: 10.1023/a:1020210615148. PMID: 12447482.

33] Lin'kova NS, Poliakova VO, Trofimov AV, Sevost'ianova NN, Kvetnoĭ IM. [Influence of peptides from pineal gland on thymus function at aging]. Adv Gerontol. 2010;23(4):543-6. Russian. PMID: 21510076.

34] Source: Khavinson, V. K., Konovalov, S. S., Yuzhakov, V. V., Popuchiev, V. V., & Kvetnoi, I. M. (2002). Modulating effects of Epithalamin and Epitalon on the functional morphology of the spleen in old pinealectomized rats. Biogerontology, 3(1), 39-45. DOI: 10.1023/a:1017989113287

35] Source: Kozina, L.S., Arutjunyan, A.V., & Khavinson, V.Kh. (2007). Antioxidant properties of geroprotective peptides of the pineal gland. Archives of Gerontology and Geriatrics, 44(2), 213-216. https://doi.org/10.1016/j.archger.2007.01.029

**FROM METABOLIC
AND ENDOCRINE ENHANCEMENT
– CHAPTER 8:**

36] Aviv Clinics, Cortisol and Cognition: How the stress hormone affects the brain, by Aaron Tribby, MS, Physiologist, November 8, 2022https://aviv-clinics.com/blog/brain-health/how-cortisol-stress-

hormone-affects-brain-
health/#:~:text=High%20chronic%20stress%20and%20
cortisol,difficult%20than%20they%20should%20be.

37] Wu J, Lin X, Huang X, Shen Y, Shan PF. Global,
regional and national burden of endocrine, metabolic,
blood and immune disorders 1990-2019: a systematic
analysis of the Global Burden of Disease study 2019.
Front Endocrinol (Lausanne). 2023 May 8;14:1101627.
doi: 10.3389/fendo.2023.1101627. PMID: 37223046;
PMCID: PMC10200867.

38] Yue X, Liu SL, Guo JN, Meng TG, Zhang XR, Li HX,
Song CY, Wang ZB, Schatten H, Sun QY, Guo XP.
Epitalon protects against post-ovulatory aging-related
damage of mouse oocytes in vitro. Aging (Albany NY).
2022 Apr 12;14(7):3191-3202. doi:
10.18632/aging.204007. Epub 2022 Apr 12. PMID:
35413689; PMCID: PMC9037278.

39] Goncharova ND, Vengerin AA, Khavinson VKh,
Lapin BA. Pineal peptides restore the age-related
disturbances in hormonal functions of the pineal gland
and the pancreas. Exp Gerontol. 2005 Jan-Feb;40(1-
2):51-7. doi: 10.1016/j.exger.2004.10.004.PMID:
15664732.

40]
https://www.targetmol.com/compound/epitalon_%28a
cetate%29

41] Korenevsky AV, Milyutina YP, Bukalyov AV,
Baranova YP, Vinogradova IA, Arutjunyan AV.
[Protective effect of melatonin and epithalon on
hypothalamic regulation of reproduction in female rats

in its premature aging model and on estrous cycles in senescent animals in various lighting regimes]. Adv Gerontol. 2013;26(2):263-274. Russian. PMID: 28976150.

42] Source: Kuznik, B. I., Pateyuk, A. V., Rusaeva, N. S., Baranchugova, L. M., & Obydenko, V. I. (2011). Effects of peptides Lys-Glu-Asp-Gly and Ala-Glu-Asp-Gly on hormonal activity and structure of the thyroid gland in hypophysectomized young chickens and old hens. Bulletin of Experimental Biology and Medicine, 150(4), 449-452. DOI: 10.1007/s10517-011-1177-3

43] N.D. Goncharova, A.A. Vengerin, V.Kh. Khavinson, B.A. Lapin, Pineal peptides restore the age-related disturbances in hormonal functions of the pineal gland and the pancreas, Experimental Gerontology, Volume 40, Issues 1–2, 2005, Pages 51-57, ISSN 0531-5565, https://doi.org/10.1016/j.exger.2004.10.004.

44] Khavinson V, Diomede F, Mironova E, Linkova N, Trofimova S, Trubiani O, Caputi S, Sinjari B. AEDG Peptide (Epitalon) Stimulates Gene Expression and Protein Synthesis during Neurogenesis: Possible Epigenetic Mechanism. Molecules. 2020 Jan 30;25(3):609. doi: 10.3390/molecules25030609. PMID: 32019204; PMCID: PMC7037223.

FROM GOOD QUALITY SLEEP
FOR IMPROVED LONGEVITY
– CHAPTER 9:

45] https://naplab.com/guides/sleep-habits-statistics/#:~:text=Most%20adults%20get%206%2D8,to%20only%2031%25%20of%20women.

46] National Heart, Lung, And Blood Institute, What Are Sleep Deprivation and Deficiency? , no author, March 24, 2022, https://www.nhlbi.nih.gov/health/sleep-deprivation

47] Stenholm, S., Head, J., Kivimäki, M., Magnusson Hanson, L. L., Pentti, J., Rod, N. H., Clark, A. J., Oksanen, T., Westerlund, H., & Vahtera, J. (2019). Sleep Duration and Sleep Disturbances as Predictors of Healthy and Chronic Disease-Free Life Expectancy Between Ages 50 and 75: A Pooled Analysis of Three Cohorts. The Journals of Gerontology: Series A, 74(2), 204-210. https://doi.org/10.1093/gerona/gly016

48] Mazzotti, D. R., Guindalini, C., Moraes, W. A. dos S., Andersen, M. L., Cendoroglo, M. S., Ramos, L. R., & Tufik, S. (2014). Human longevity is associated with regular sleep patterns, maintenance of slow wave sleep, and favorable lipid profile. Frontiers in Aging Neuroscience, 6(134). https://doi.org/10.3389/fnagi.2014.00134

49] Hou, C., Lin, Y., Zimmer, Z., Tse, L. A., & Fang, X. (2020). Association of sleep duration with risk of all-cause mortality and poor quality of dying in oldest-old people: a community-based longitudinal study. BMC

Geriatrics, 20(357). https://doi.org/10.1186/s12877-020-01759-6

50] Arutjunyan, A., Kozina, L., Milyutina, Y., Korenevsky, A., Stepanov, M., & Arutyunov, V. (2012). Melatonin and pineal gland peptides are able to correct the impairment of reproductive cycles in rats. Current Aging Science, 5(3), 178-185. https://doi.org/10.2174/1874609811205030003

51] Khavinson, V., Goncharova, N., & Lapin, B. (2001). Synthetic tetrapeptide epitalon restores disturbed neuroendocrine regulation in senescent monkeys. Neuro Endocrinology Letters, 22(4), 251-254.

52] Goncharova, N. D., Vengerin, A. A., Khavinson, V. Kh., & Lapin, B. A. (2005). Pineal peptides restore the age-related disturbances in hormonal functions of the pineal gland and the pancreas. Experimental Gerontology, 40(1-2), 51-57. https://doi.org/10.1016/j.exger.2004.10.004

53] Yale School of Medicine, Sleep's Crucial Role in Preserving Memory, by Jordan Sisson, May 10, 2022, https://medicine.yale.edu/news-article/sleeps-crucial-role-in-preserving-memory/

54] Oyetakin-White P, Suggs A, Koo B, Matsui MS, Yarosh D, Cooper KD, Baron ED. Does poor sleep quality affect skin ageing? Clin Exp Dermatol. 2015 Jan;40(1):17-22. doi: 10.1111/ced.12455. Epub 2014 Sep 30. PMID: 25266053.

FROM SKIN HEALTH AND EPITALON – CHAPTER 10:

55] American Academy of Dermatology Association, www.aad.org, Skin Cancer, Incident Rates, no author or year of publication, https://www.aad.org/media/stats-skin-cancer

56] Yakupu, A., Aimaier, R., Yuan, B., Chen, B., Cheng, J., Zhao, Y., Peng, Y., Dong, J., & Lu, S. (2023). The burden of skin and subcutaneous diseases: findings from the global burden of disease study 2019. Frontiers in Public Health, 11, 1145513. https://doi.org/10.3389/fpubh.2023.1145513

57] Bocheva G, Slominski RM, Slominski AT. Environmental Air Pollutants Affecting Skin Functions with Systemic Implications. Int J Mol Sci. 2023 Jun 22;24(13):10502. doi: 10.3390/ijms241310502. PMID: 37445680; PMCID: PMC10341863.

58] Khavinson VK, Popovich IG, Linkova NS, Mironova ES, Ilina AR. Peptide Regulation of Gene Expression: A Systematic Review. Molecules. 2021 Nov 22;26(22):7053. doi: 10.3390/molecules26227053. PMID: 34834147; PMCID: PMC8619776.

59] Gutop, E.O., Linkova, N.S., Kozhevnikova, E.O. et al. AEDG Peptide Prevents Oxidative Stress in the Model of Induced Aging of Skin Fibroblasts. Adv Gerontol 12, 143–148 (2022). https://doi.org/10.1134/S2079057022020096

60] Khavinson V, Diomede F, Mironova E, Linkova N, Trofimova S, Trubiani O, Caputi S, Sinjari B. AEDG Peptide (Epitalon) Stimulates Gene Expression and

Protein Synthesis during Neurogenesis: Possible Epigenetic Mechanism. Molecules. 2020 Jan 30;25(3):609. doi: 10.3390/molecules25030609. PMID: 32019204; PMCID: PMC7037223.

61] Chalisova, N. I., Lin'kova, N. S., Zhekalov, A. N., Orlova, A. O., Ryzhak, G. A., & Khavinson, V. Kh. (2014). Short peptides stimulate skin cell regeneration during aging. Adv Gerontol, 27(4), 699-703.

62] Vinogradova, I. A., Bukalev, A. V., Zabezhinski, M. A., Semenchenko, A. V., Khavinson, V. Kh., & Anisimov, V. N. (2008). Geroprotective effect of Ala-Glu-Asp-Gly peptide in male rats exposed to different illumination regimens. Bulletin of Experimental Biology and Medicine, 145, 472–477.

63] Gutop, E. O., Linkova, N. S., Kozhevnikova, E. O., Fridman, N. V., Ivko, O. M., & Khavinson, V. Kh. (2022). AEDG Peptide Prevents Oxidative Stress in the Model of Induced Aging of Skin Fibroblasts. Advances in Gerontology, 12, 143-148.

64] Avolio, F., Martinotti, S., Khavinson, V. Kh., Esposito, J. E., Giambuzzi, G., Marino, A., Mironova, E., Pulcini, R., Robuffo, I., Bologna, G., Simeone, P., Lanuti, P., Guarnieri, S., Trofimova, S., Procopio, A. D., & Toniato, E. (n.d.). Peptides Regulating Proliferative Activity and Inflammatory Pathways in the Monocyte/Macrophage

FROM VISION AND OCULAR PROTECTION – CHAPTER 11:

65] World Health Organization, Blindness and vision impairment, no author, 10 August 2023, https://www.who.int/news-room/fact-sheets/detail/blindness-and-visual-impairment#:~:text=Key%20facts,are%20refractive%20errors%20and%20cataracts

66] Kozina, L. S., Arutjunyan, A. V., & Khavinson, V. Kh. (2007). Antioxidant properties of geroprotective peptides of the pineal gland. Archives of Gerontology and Geriatrics, 44(Supplement), 213-216. https://doi.org/10.1016/j.archger.2007.01.029

67] Khavinson, V., Diomede, F., Mironova, E., Linkova, N., Trofimova, S., Trubiani, O., Caputi, S., & Sinjari, B. (2020). AEDG Peptide (Epitalon) Stimulates Gene Expression and Protein Synthesis during Neurogenesis: Possible Epigenetic Mechanism. Molecules, 25(3), 609. https://doi.org/10.3390/molecules25030609

68] Khavinson, V., Diomede, F., Mironova, E., Linkova, N., Trofimova, S., Trubiani, O., Caputi, S., & Sinjari, B. (2020). AEDG Peptide (Epitalon) Stimulates Gene Expression and Protein Synthesis during Neurogenesis: Possible Epigenetic Mechanism. Molecules, 25(3), 609. https://doi.org/10.3390/molecules25030609

69] Egorov, V. V., Sorokin, E. L., & Smoliakova, G. P. (2003). Administration of Epithalamine in the treatment of unstable glaucoma of different types after normalization of intraocular pressure. Vestnik

Oftalmologii, 119(1), 5-7. PMID: 12608033. Retrieved from https://europepmc.org/article/med/12608033

70] Khavinson, V., Razumovsky, M., Trofimova, S., Grigorian, R., & Razumovskaya, A. (2002). Pineal-regulating tetrapeptide epitalon improves eye retina condition in retinitis pigmentosa. Neuroendocrinology Letters, 23(5), 365-368.

71] Trofimova, S. V., & Khavinson, V. K. (2001). [Effectiveness of bio-regulators in the treatment of diabetic retinopathy]. Vestnik Oftalmologii, 117(3), 11-14.

FROM BONE AND JOINT HEALTH – CHAPTER 12:

72] Centers for Disease Control And Prevention, Arthritis Related Statistics, no author, October 4, 2023, https://www.cdc.gov/arthritis/data_statistics/arthritis-related-stats.htm#:~:text=National%20Prevalence,gout%2C%20l upus%2C%20or%20fibromyalgia.

73] American Medical Wellness, Best Peptides For Longevity, no author, December 1, 2023, https://www.americanmedicalwellness.com/best-peptides-for-longevity/

74] Khavinson, V., Diomede, F., Mironova, E., Linkova, N., Trofimova, S., Trubiani, O., ... & Sinjari, B. (2020). AEDG Peptide (Epitalon) stimulates gene expression and protein synthesis during neurogenesis: possible epigenetic mechanism. Molecules, 25(3), 609.

75] Eden Aesthetics, Anti-Aging and Reverse Aging Therapy: NAD+ / NMN / Epitalon / Stem Cells, no author, Dec 6, 2023, https://www.edenderma.com/post/anti-aging-and-reverse-aging-therapy-nad-epitalon-stem-cells

76] South Lake Pharmacy, Anti-Aging/Rejuvinating, https://sa1s3.patientpop.com/assets/docs/394982.pdf

77] Anisimov, V. N., Khavinson, V. K., Popovich, I. G., Zabezhinski, M. A., Alimova, I. N., Rosenfeld, S. V., Zavarzina, N. Y., Semenchenko, A. V., & Yashin, A. I. (2003). Effect of Epitalon on biomarkers of aging, life span and spontaneous tumor incidence in female Swiss-derived SHR mice. Biogerontology, 4(4), 193-202. https://doi.org/10.1023/a:1025114230714

78] Al-Qadhi, G., Aboushady, I., & Al-Sharabi, N. (2021). The Gingiva from the Tissue Surrounding the Bone to the Tissue Regenerating the Bone: A Systematic Review of the Osteogenic Capacity of Gingival Mesenchymal Stem Cells in Preclinical Studies. Translational Applications of Tissue Engineering and Regenerative Medicine, 2021, Article ID 6698100. https://doi.org/10.1155/2021/6698100

FROM DOSAGE AND ADMINISTRATRION – CHAPTER 13:

79] Good Laboratory Practice Bioscience, Epitalon TFA, https://www.glpbio.com/epitalon-tfa.html

80] Yue, X., Liu, S. L., Guo, J. N., Meng, T. G., Zhang, X. R., Li, H. X., ... & Guo, X. P. (2022). Epitalon protects against post-ovulatory aging-related damage of mouse oocytes in vitro. Aging (Albany NY), 14(7), 3191.

81] Khavinson, V. K., Izmaylov, D. M., Obukhova, L. K., & Malinin, V. V. (2000). Effect of epitalon on the lifespan increase in Drosophila melanogaster. Mechanisms of ageing and development, 120(1-3), 141-149.

82] Khavinson, V., Diomede, F., Mironova, E., Linkova, N., Trofimova, S., Trubiani, O., ... & Sinjari, B. (2020). AEDG Peptide (Epitalon) stimulates gene expression and protein synthesis during neurogenesis: possible epigenetic mechanism. Molecules, 25(3), 609.

83] Khavinson, V. K., Bondarev, I. E., & Butyugov, A. A. (2003). Epithalon peptide induces telomerase activity and telomere elongation in human somatic cells. Bulletin of Experimental Biology and Medicine, 135(6), 590-592.

84] Asinov Anisimov, V. N., Khavinson, V. K., Popovich, I. G., Zabezhinski, M. A., Alimova, I. N., Rosenfeld, S. V., Zavarzina, N. Y., Semenchenko, A. V., & Yashin, A. I. (2003). Effect of Epitalon on biomarkers of aging, life span and spontaneous tumor incidence in female Swiss-derived SHR mice. Biogerontology, 4(4), 193-202. https://doi.org/10.1023/a:1025114230714

85] Alzheimer's Drug Discover Foundation, Cognitive Vitality.org, Epithalamin, Epithalon, https://www.alzdiscovery.org/uploads/cognitive_vitality_media/Epithalamin-and-Epithalon-Cognitive-Vitality-For-Researchers.pdf

FROM CASE STUDIES
AND PERSONAL EXPERIENCES
– CHAPTER 14:

86] Anisimov, V. N., Khavinson, V. K., Popovich, I. G., Zabezhinski, M. A., Alimova, I. N., Rosenfeld, S. V., Zavarzina, N. Y., Semenchenko, A. V., & Yashin, A. I. (2003). Effect of Epitalon on biomarkers of aging, life span and spontaneous tumor incidence in female Swiss-derived SHR mice. Biogerontology, 4(4), 193-202. https://doi.org/10.1023/a:1025114230714

87] Trofimova, S. (2020). Molecular Mechanisms of Retina Pathology and Ways of Its Correction. Springer International Publishing.

88] Trofimova, S., & Trofimova, S. (2020). Results of the Clinical Study of Short Peptides (Cytogens) in Ophthalmology. *Molecular Mechanisms of Retina Pathology and Ways of its Correction*, 69-84.

89] Khavinson, V., Diomede, F., Mironova, E., Linkova, N., Trofimova, S., Trubiani, O., Caputi, S., & Sinjari, B. (2020). AEDG Peptide (Epitalon) Stimulates Gene Expression and Protein Synthesis during Neurogenesis: Possible Epigenetic Mechanism. Molecules, 25(3), 609. https://doi.org/10.3390/molecules25030609

90] Yue, X., Liu, S. L., Guo, J. N., Meng, T. G., Zhang, X. R., Li, H. X., ... & Guo, X. P. (2022). Epitalon protects against post-ovulatory aging-related damage of mouse oocytes in vitro. Aging (Albany NY), 14(7), 3191.

91] Goncharova, N. D., Vengerin, A. A., Khavinson, V. K., & Lapin, B. A. (2005). Pineal peptides restore the age-related disturbances in hormonal functions of the pineal gland and the pancreas. Experimental gerontology, 40(1-2), 51-57.

FROM THE FUTURE OF EPITALON IN ANTI-AGING
– CHAPTER 15:

92] Khavinson, V., Diomede, F., Mironova, E., Linkova, N., Trofimova, S., Trubiani, O., Caputi, S., & Sinjari, B. (2020). AEDG Peptide (Epitalon) Stimulates Gene Expression and Protein Synthesis during Neurogenesis: Possible Epigenetic Mechanism. Molecules (Basel, Switzerland), 25(3), 609. https://doi.org/10.3390/molecules25030609

93] Anisimov VN, Khavinsov VKh, Alimova IN, Provintsiali M, Manchini R, Francheski K. Epithalon inhibits tumor growth and expression of HER-2/neu oncogene in breast tumors in transgenic mice characterized by accelerated aging. Bull Exp Biol Med. 2002 Feb;133(2):167-70. doi: 10.1023/a:1015555023692. PMID: 12428286.

94] Anisimov VN, Khavinson VKh, Popovich IG, Zabezhinski MA. Inhibitory effect of peptide Epitalon on colon carcinogenesis induced by 1,2-dimethylhydrazine in rats. Cancer Lett. 2002 Sep 8;183(1):1-8. doi: 10.1016/s0304-3835(02)00090-3. PMID: 12049808.

95] Knox, E. G., Aburto, M. R., Clarke, G., Cryan, J. F., & O'Driscoll, C. M. (2022). The blood-brain barrier in aging and neurodegeneration. Molecular psychiatry, 27(6), 2659–2673. https://doi.org/10.1038/s41380-022-01511-z

96] Sibarov DA, Vol'nova AB, Frolov DS, Nosdrachev AD. [Intranasal epitalon infusion modulates neuronal activity in the rat neocortex]. Ross Fiziol Zh Im I M Sechenova. 2006 Aug;92(8):949-56. Russian. PMID: 17217245.

97] Khavinson VKh, Linkova NS, Kvetnoy IM, Kvetnaia TV, Polyakova VO, Korf HW. Molecular cellular mechanisms of peptide regulation of melatonin synthesis in pinealocyte culture. Bull Exp Biol Med. 2012 Jun;153(2):255-8. English, Russian. doi: 10.1007/s10517-012-1689-5. PMID: 22816096.

98] Apostolopoulos V, Bojarska J, Chai TT, Elnagdy S, Kaczmarek K, Matsoukas J, New R, Parang K, Lopez OP, Parhiz H, Perera CO, Pickholz M, Remko M, Saviano M, Skwarczynski M, Tang Y, Wolf WM, Yoshiya T, Zabrocki J, Zielenkiewicz P, AlKhazindar M, Barriga V, Kelaidonis K, Sarasia EM, Toth I. A Global Review on Short Peptides: Frontiers and Perspectives. Molecules. 2021 Jan 15;26(2):430. doi: 10.3390/molecules26020430. PMID: 33467522; PMCID: PMC7830668.

Made in the USA
Columbia, SC
26 September 2024

43094285R00130